PLACE NAMES *and* FIELD NAMES *of*
NORTHUMBERLAND

PLACE NAMES *and* FIELD NAMES *of*
NORTHUMBERLAND

STAN BECKENSALL

TEMPUS

M. L. Leeds
79 Upper North Street
Brighton
East Sussex
BN1 3FL

First published 2006

Tempus Publishing Limited
The Mill, Brimscombe Port,
Stroud, Gloucestershire, GL5 2QG
www.tempus-publishing.com

© Stan Beckensall, 2006

The right of Stan Beckensall to be identified as the Author
of this work has been asserted in accordance with the
Copyrights, Designs and Patents Act 1988.

All rights reserved. No part of this book may be reprinted
or reproduced or utilised in any form or by any electronic,
mechanical or other means, now known or hereafter invented,
including photocopying and recording, or in any information
storage or retrieval system, without the permission in writing
from the Publishers.

British Library Cataloguing in Publication Data.
A catalogue record for this book is available from the British Library.

ISBN 0 7524 3647 3

Typesetting and origination by Tempus Publishing Limited
Printed in Great Britain

CONTENTS

Preface	7
Acknowledgements	7
Introduction	8
PART I PLACE NAMES	15
The setting	17
Settlers from northern Europe	20
Place name elements	22
Map locating places on the grid	48
Alphabetical list of place names in Northumberland	49
PART II FIELD NAMES	75
General review	77
Alphabetical list of common elements (denominatives) in field names	83
A concentrated area of study: Thirston, Felton and Eshott	86
Acklington	96
Shilbottle and Brainshaugh	100
Low Buston and Warkworth	102
Lesbury, Bilton and Longhoughton	107
Rennington and Fallodon	113
Ellingham and Tughall	115
Newham, Newstead and Rosebrough	117
Chatton and Lyham	118
Lucker and Detchant	121
Scremerston, Spindlestone, Outchester and Bamburgh	121
Pawston and Alnham	124
Elsdon, Monkridge, Woodside, Troughend, Otterburn and Rochester	125
South Northumberland, with Newlands and Whittonstall, Dilston, Throckley and Coastley	128
Alphabetical list of field names	134
Bibliography	159

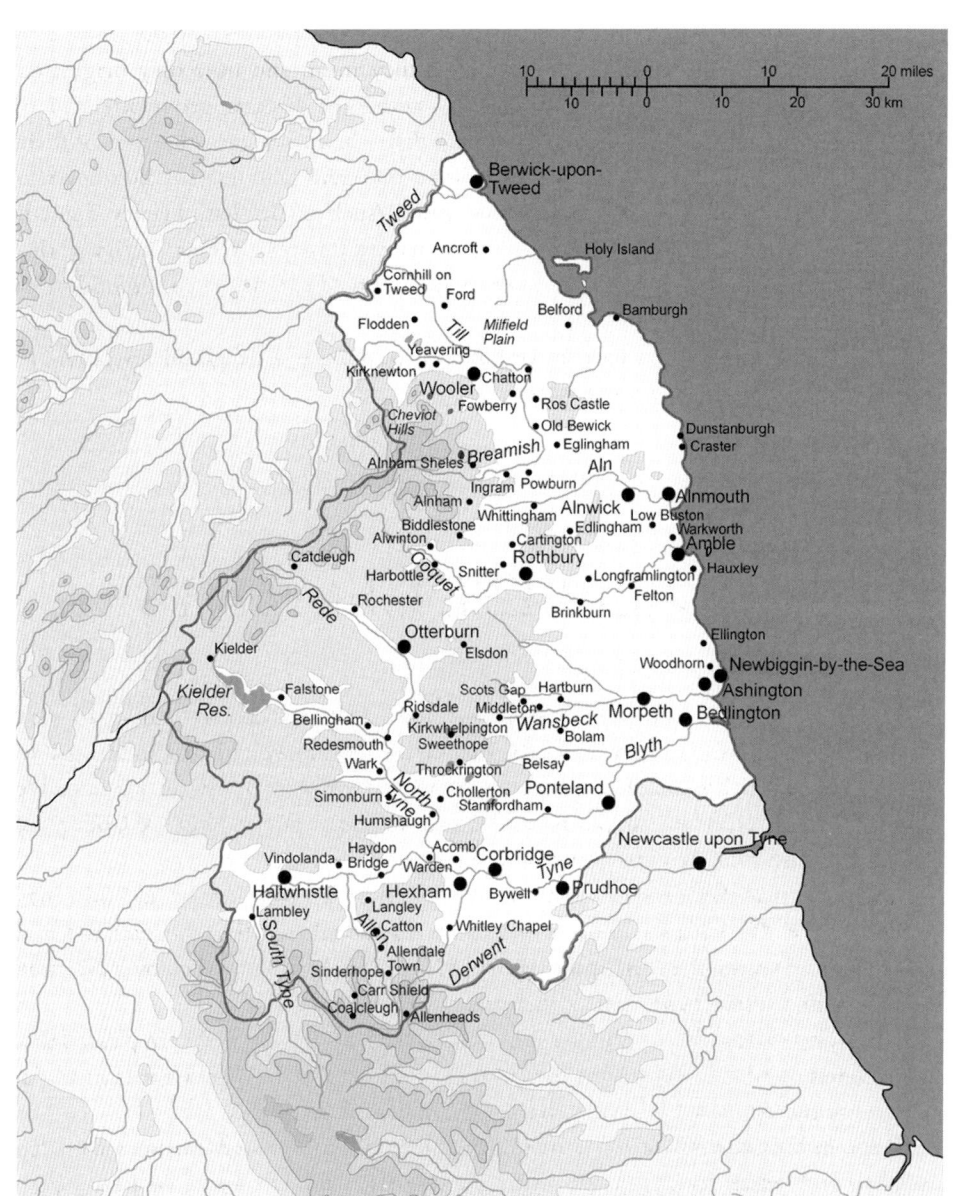

1 Map of Northumberland featuring significant place names. *Copyright Marc Johnstone*

PREFACE

Although the names of places where we live are an integral part of our lives, it is surprising how little attention is given to what they mean and how they originated. They have been notably absent from many studies of history, geography and/or environmental studies in our schools for so long that when I became conscious of them I felt compelled to share my findings. In my Stoke-on-Trent schooldays it did not occur to my teachers to tell us what place names meant (such as Burslem, Hanley, Tunstall, Longton and Mow Cop), or why Etruria was so named. I may have been intrigued that some of the lads at my high school were called Tunnicliffe, Greatbatch, Proudfoot or Bracegirdle, but that was all.

To discover that all names have an origin and meaning is a delightful revelation; to have no curiosity about them is a minor tragedy.

ACKNOWLEDGEMENTS

In the 1970s the Duke of Northumberland and his staff allowed access to, and help with, the considerable archive at Alnwick Castle that made the section on field names possible. Terry White of Felton photographed the Robert Norton maps with considerable professional skill. Richard Parkin, then a lecturer in geography at Alnwick College of Education, converted the 6-inch maps that I compiled into works of art.

I owe a particular debt of gratitude to Heather Wilson for typing the 1975 *Northumberland Place Names* book and to Ann Markwick for the concentrated and difficult task of typing the *Northumberland Field Names* book. Having to go over the same ground again, I realise how difficult that was.

I have acknowledged help from farmers and students in the text, and I am pleased that their interest continues. I am also grateful to the County Record Office for their initial help in my surveys. Marc Johnstone has provided the location map.

All this information will be of value to the planned Great North Museum, which can use place and field names as an important component in the history of the North East.

I am happy to share this work freely with anyone who wishes to pursue it, provided that it is acknowledged. The scope for further research is enormous.

INTRODUCTION

Names spread over a map have an immediate attraction. The sounds are interesting: the softness of Wooler, the abruptness of Ford, the rhythmic flow of Bellingham give music to the landscape. One must know that in Bellingham the latter part is pronounced 'injum', which applies to most others with the same ending except Chillingham, with a hard g. It is a local characteristic with no really satisfactory explanation. The incomer may find difficulty with Ulgham, pronounced 'uffm', and may want to know why. this is the case. Although Ashington, once a mining town, may appear to be explained by industrial waste, and one might be forgiven for thinking that Haltwhistle is accounted for because it has a railway station, we have to look below the surface to find the true meaning. Alnwick, now famous for its Harry Potter connection, its garden and tree house, still gives people problems with its pronunciation, but 'alln-wick' has to be abandoned for 'ann-ick'. A scatter of these -wick endings extends from Berwick-on-Tweed through Cheswick, Goswick, Lowick and Bewick. The reason for this has to come from an examination of documents which show earliest spellings and changes.

Many people take their names from places: Windsor, London, Hampshire and Becconsall spring to mind; locally there are Charltons, Fenwicks and Bewicks, for example. Street names preserve some of our history: Priestpopple and St Cuthbert's Terrace; Bede and Wilfrid recall a 'Golden Age' of Northumbrian Christianity. More recently, the names of local worthies name streets. The miners' leader Pete Leatherland, later to train as a teacher at Alnwick College, is remembered at Shilbottle; Jack Pescott of Hexham also gives his name to a street.

It is, however, the names of fields that give the most graphic account of what happened in the past, for most of Northumberland's history and wealth was based on agriculture. Before the level of mechanisation that we have reached today cut down the number of agricultural workers, most people worked on the land. They gave fields their names – sometimes their own, sometimes naming them after landscape features, crops, fertile and poor areas, animals and vegetation, proximity to springs or roads, and other features. They added their sense of humour, too. Field names are intensely personalised, and often names would be erased and replaced when a farm changed hands.

Language is not just about the communication of meaning, but of feeling also. Karenza Storey of Harbottle chose these field names from my 1977 study to create a poem:

Job's Northumberland Fields
Jobs Field, Jockeys Field, Juniper Haugh Waste,
Obelisk, Out Peece, Ottercops,
Blowbutts, Bakethin, Birdhopecraig,
Spindle Burn, Shovelbred, Saltgrese;

Foul Brigges, Farny Knowle, Featherblow,
Imberley, Improvement, Irish Dales,
Evenstones, Elyhaughe, Estnoyke,
Light Pipe, Ligger, Labour in Vain,
Divshey Dyke, Doe Path, Dirty Doup,
Sweeting Roods, Sunny Close, Swanlawflat

(The poem first appeared in *KEY words*, Ed. A. Stevenson, Vol. 1 no. 4, 1985, MidNAG)

The names of fields change with time much more than those of villages and towns. Often we know about field names by the accidental retention of documents or legal retention in estate and other offices, or at farms themselves, but many must have disappeared from the record.

Names are made up of elements. Thus Blyth has one sound, Bywell has two, and Chillingham has three. There may be an addition: Kirknewton has 'church' added to 'new settlement', Longframlington is drawn out along a main street. 'New' and 'old', 'upper' and 'lower' show an expansion. So one way to begin a study of names is to separate them into their parts, or elements.

The common endings of place names include : -ton, -ham, -wick. A glance at the map will show that -ing has been put before -ton and -ham in places like Eglingham and Whittington. Wick is often put after another element, as in Prendwick.

Why so many of these elements? What do they mean? These elements are very old, and are present all over Britain (Birmingham, Brighton, Prestwick, for example). As we look further for similarities and differences in names on a map of Northumberland, we may notice that Bamburgh and Dunstanburgh have a similar element to Edinburgh. There are many chesters; two of them are Roman forts on Hadrian's Wall. Is there a connection between Rothbury and Bury in Lancashire, and do all Prestons have the same meaning?

As a teacher, I used to ask my younger pupils to draw for themselves an island, put in any features they wanted, and then add names. This became the basis for stories which they would compose later. What emerged was physical features: mountains, hills, rivers, streams, inlets, cliffs – the first to be named. If someone decided to add a bridge, the word itself was not considered interesting enough by itself; it needed a name to distinguish it from other bridges. Woodland, farmlands and settlements also needed to be named, but how? There came a time when, as stories began to appear, so did new names to mark incidents. Those who made the island tropical were likely to have buried treasure, perhaps cannibals and pirates. This is an ancient process of wanting to socialise the landscape, to name its parts, to create it in a way in our own image. Rather like the

2 Landscape: Cheviot hills

story of Adam and Eve when they had to name the animals, and I wonder what the story-tellers might have added had they extended this to different parts of Eden.

Just as geology ('solid' underneath 'surface') determines conditions for survival in its relief, types of rock (coal-bearing, and metal-bearing, for example), fertility of soil, drainage, and what can grow there, so settlement patterns are dependent on all these factors. We know much about the earliest people who settled Northumberland, even though they did not leave a language, from archaeology. With the development of writing there was a chance of our knowing how they gave the land its names, and it is an accumulation of documents that enables us to make this study. However, the recorded names must always be examined against what is actually in the landscape in order to make sense. Field systems, enclosures, defences and buildings are there to be 'read' and interpreted in our early history, even without language.

The first fragmentary recording of what existed came with the Roman occupation in the first century AD, when the Romans found a well-developed agricultural and pastoral society with its tribal power-structure in place. We know that European tribes had their own languages, and soldiers in the Roman army who were recruited from these tribes must have continued to use them. However, in the interests of uniformity and efficiency all were grounded in some basic Latin. The Vindolanda writing tablets give an insight into the kind of written communication used in the business, military and social world of that garrison.

Conquerors may adapt place names already in existence rather than choose entirely new names, though both processes tend to go together. The Romans 'Latinised' some of them, but did not leave much evidence on our maps. Latin came through our language largely through Norman French with William the Conqueror, and forms its major component. The rest, Anglo-Saxon or Old English, however, has the greatest impact on place names, in terms of elements retained by the Normans.

As this book progresses these developments will be traced in more detail, but to end this introduction, place names already given in the text will be chosen to show how the process works. In each case the earliest spelling and changes show, to some extent, how the names came into being:

Wooler was *Wullour* in 1187, then *Welloure, Wllovera,* and *Wullour.* One element is *ofre,* a rare one, with its sound being close to 'over'. The 'well' part is water, usually a spring, so this name probably fits its position on the edge of high ground looking over the water to the Milfield Plain. Both elements are Old English.
Ford, a common name at river crossings, was *Forda* in 1224.
Bellingham was *Bainlingham* in 1170 and became Bellingham in 1254. The meaning is not certain. 'Bell' is a word for a hill, but it is also a personal name. The 'ingham' part is two elements, meaning 'the settlement associated with a named person', so it could have been the family of Bel settling there or the place of the hill-dwellers.
Chillingham was *Cheuligeham* in 1186, and *Chevelingham* in 1231. It appears as Chillyngham in 1348. Again it is named after a settler, this time *Ceofel.*
Ulgham was *Wlacam* in 1139, *Ulweham* in 1242, *Ulcham* in 1251, and *Ulgham* in 1290. There is also an Ulwham at Featherstone, much further to the west. Old English *ule* is an owl, and *hwamm* is a corner or angle, so this is a valley, nook or other place where owls nested. Notice that it is not *ham.*

3 Signpost from the Otterburn area

Haltwhistle was *Hautwisel* in 1240, and *Hawtewysill* in 1279. The Old English *heafod* is a hill, and *twisla* is a fork in the river. This fits its position on high ground above water. (Twizel bridge near Ford, which the English army crossed on the way to Flodden in 1513, is the place where the rivers Tweed and Till meet).
Alnwick, like many other settlements on the River Aln, takes its name from a British river name recorded by Ptolemy in c.150 and by Bede in c.730 as *Alaunos* and *Alauna*, but it is not clear what it means.
Berwick, again with a second element marking it as a farm, was *Berewich* in 1167, and, like many others with the same element, means that the farm specialised in growing barley.
Cheswick (pronounced chizic), was *Chesewic* in 1208, and means cheese-farm.
Goswick, (pronounced gozic), was *Gossewic* in 1202, goose-farm.
Lowick, *Lowich* in 1181, is a farm on the River Low, actually a stream rather than a river, where a *low* in dialect is a shallow pool, usually left by the retreating tide.
Bewick (Old and New), was *Bowich* in 1167, and *Bewick* in 1296, from Old English *beo*, bee-farm.
Charlton, common in England, *Carlton* in 1195, is the settlement of the free peasant.
Fenwick, near Kyloe, is *Fenwic* in 1208, like another near Stamfordham, and means the farm on the fen.
Priestpopple in Hexham is not a 'place' name but part of the town where priests held small parcels of land. This element occurs frequently in field names.
Shilbottle was *Siplibotle* in 1228, and *Schiplibotle* in 1238, becoming *Schiplingbothill* in 1242. The element *botl* is Old English, meaning a building, a dwelling. The people who first lived there were probably called Shipley.
Hexham is a complex name that hides its origins. In 681 at the time when the great church there was founded by Wilfrid, it appeared as *Hagustaldes ea* and four years later as *Hagustaldes ham*. It changes to Hexteldesham in 1188, and appears as Hexham from 1362. The Old English elements of *ea* and *ham* denote a settlement by water (Hexham being on the Tyne). The *hagustald* was the son of a person of some note who had to go and find land for himself because he could not inherit much more than a smallholding. Early spellings (with *eg* or *ei*) also point to an island of good land in moorland.
Harbottle, *Hirbotl* around 1220 and *Herbotle* in 1291, was already the site of a castle when it was named, the building referred to presumably in *botl*. It was Herbotill in 1539. In Old English a *hyra* is someone who is hired, probably referring to people in the army.
The river Blyth was *Blitha* in 1204, meaning that it was a gentle, merry or pleasant river. The town name appeared earlier in 1130 as *Blida*.
Bywell, *Biguell* in 1104, and *Biewell* in 1195, lies on a bend of the Tyne, so the Old English elements *byge-wella* confirm that it is a spring in the bend of that river.
Kirknewton, on the edge of the Cheviots in the Glen valley, has two very common elements that mean a new settlement, with the 'kirk' added to show it has a church, to distinguish it from others.
Longframlington began its literary life as *Fremelintun* in 1166. 'Ing-ton' is associated with someone called Fram or Framela. It stretches along a modern main road and along an old

Roman road. Eglingham belongs to a similar run of names in which the personal names of founders determine the difference (Edlingham, Ellingham). It was *Ecgwulfincham* in 1050, with the founder's name given as *Ecgwulf*.

Whittington (Great and Little) was *Witynton* in 1233, and was named after Hwita.

Prendwick was *Prendewic* in 1242 and *Prandewick* in 1279. It looks like Prend's farm.

Bamburgh, one of the most famous places in Northumberland's history, is one of the earliest to be documented. In the Anglo Saxon Chronicle it appears as *Bebbanburh*, and in Bede in 890 as *Baenburg*. It is Bebbe's or Bebban's burgh, the fortification of Bebba (e), queen of Aethelfrith (593-617).

Dunstanburgh is a three element name, but the Dunstan recorded in 1242 with the same spelling means the stone hill. The hill is in fact an outcrop of whinstone, where boiling rock has crystallised to form pillars of basalt into a promontory and a strong point, used for the building of a castle. It is possible that an earlier fortification lay there; at Bamburgh too. Thus the common element 'burg', a 'dun' being a hill.

Chesters has an unusual prefix to the 'castra' element, meaning a fort: it was *Scytlecester* in 1104-8, which may be a personal name but is more likely to be from *scyttels* meaning an enclosure. It was not unusual for old walled enclosures to be used for animals.

Rothbury was *Routhebiria* in 1125, *Rodebir* in 1271, and *Routhebyr* in 1291. It may be the *burh*, the fortification, of Hrotha, but another meaning is that it could be the 'red fort' on the Coquet, perhaps coloured red by iron or crushed porphyry.

Finally, Preston, a common place name, is the farm or settlement of the priests (Old English *Preosta-tun*).

4 Dunstanburgh Castle from the south

What is obvious from this sample is that most early forms of the names are not very ancient. After William the Conqueror had devastated the North there was little left to record by scribes compiling the Domesday Book, the record of what he had gained. The writings of those such as Bede in the monasteries were the earliest records, then there is a gap.

The other notable feature is that all the elements are Old English. The Normans stuck approximately to place names as they heard them from local people. French names are rare.

Another feature is that some names include people who have disappeared without trace from other written records, and that others refer to landscape and animals.

PART I

PLACE NAMES

THE SETTING

To understand how and why places get their names, we need to look carefully at the landscapes, formed by geology and by the impact of people who have exploited the land for thousands of years.

Northumberland has a long coastline, hills formed by volcanic action, and a series of rock formations that include sandstone, limestone, shale and coal, with an intrusion of dykes of basalt, known locally as whinstone. To the north-west are the volcanic Cheviot Hills, from which spread sedimentary rocks, of varying height, including scarps of sandstone. Although there is coal, lime and iron almost everywhere, the great coal seams lie in the south-east, the basis of the Industrial Revolution. Galena, the basis of lead smelting, lies in limestone to the south. Valleys, coastal plains and the gravels, such as those of the Milfield Plain, grow the best crops, but much high ground has a lighter, more acidic covering of soil, some producing heather, bracken and coarse grass. Thin soils have recently been planted with timber as a cash crop.

What we see today is the result of people using the land for centuries, and changing it in both dramatic and subtle ways, so when we are looking at the time when place names were given and written down, we have to take these changes into account.

The present landscape was established in prehistoric times, with mixed farming. Hunting played a great part always, and so did the management of natural vegetation, but the slow introduction of arable agriculture changed ways of life, for people became more attached to one spot, where fields, homes and enclosures were built. By the time the Romans came, this agricultural settlement was well established, and society was organised with its leaders or 'hierarchies'. We can still see some of these early field systems in the landscape and from the air, although many lie buried under later land-use, particularly where the fertility of the land has declined and growing crops gave way to grassland for pasture, and 'waste' or moorland. These are the places where few people want to live today in any numbers, such as in the Cheviot Hills, and that is where some of the ancient field systems of terraces, lynchets, cord rig and s-bends of rig and furrow are to be seen, especially when the sun is low in the sky; some are prehistoric.

River valleys and plains have always been attractive places to settle, as the soil is generally easy to work and well-drained, and it is interesting that river names tend to retain their earliest pre-Roman ('British') forms. For people coming in by sea, river estuaries and river valleys would have been good entry points to push further inland. Almost all Northumberland rivers flow from west to east to the sea. The Breamish/Till is an exception, turning north after it leaves the Cheviots at Powburn, running north, cutting north-west through the sandstone scarp at Weetwood Bridge, then flowing north again to meet the Tweed.

The landscape is varied, with scarps, fault lines, and many minor undulations that provide sheltered places and ridges with extensive views. The land rises on the edges of the Pennines and becomes bleaker and colder, with sparse settlement at any time, except for mining and smelting areas. Archaeology has revealed buried soils that tell of minor climate changes and different degrees of fertility. Our knowledge of what happened in the past is always fragmentary, but there is sufficient evidence to form an outline.

There is no prehistoric language known to us, and some of our knowledge of early people living in Northumberland, though biased, comes to us from the Roman settlers and writers. We know, for example, that the two large northern tribes in the area are named Brigantes and Votadini. Hadrian's Wall marks the halt of the Empire, a line forced between civilisation and the 'Britunculi' (named in the Vindolanda writing tablets), those who lived north of the Wall. It was not a line of demarcation between England and Scotland, as neither existed then.

The land was conquered, but as the Roman occupation lasted for so long, we must not regard it as a single phenomenon. Although the army was provisioned from various supply bases along over 62 miles of frontier, by a system of roads and harbours, it is likely that local people were in a position to trade products such as grain, livestock and hunted animals with them. It is also likely that they would come to learn something of the occupiers' language and vice versa. As Roman troops were widely recruited it is even possible that communication could take place in a different tribal language.

Despite the hundreds of years that passed, there is hardly anything of this occupation alluded to in the place names. The archaeological evidence shows a number of upland farms with traditional round huts in use, when the Romans favoured rectangular and square buildings. Little in the way of traded goods seems to be in the record, so perhaps people were left to go on living in their traditional way provided that they behaved themselves. Very little of the modern county would have been under direct occupation, and outlying areas would probably have been patrolled.

So what is left of that span of over 300 years? The main feature is the Wall and all the military and civil buildings that were attached to it, whether close by or distant. Of the names, the English Heritage handbook, for example, lists them as Walltown Crags, Cawfields, Whinshields, Housesteads, Sewingshields, the Temple of Mithras at Carrawburgh, Black Carts Turret, Great Chesters, Little Chesters, Brunton Turret, and Heddon on the Wall. There are others, such as Carvoran, and Halton. When one adds Wall village and others to the list, then takes into account the important settlements south of the Wall, such as Vindolanda and Corbridge, there is still very little Roman influence in the names, and we must dig deeper to find something older. The 'Wall' recurs in names, and the element 'chesters' is from 'castra'; the 'burgh' is an Old English name for a fortification.

Outlying forts like Rochester carry the 'castra' element; The Stanegate, a medieval name, behind the Wall, the same as Stane Street in meaning, refers to the hard surface of that early frontier. There is a Watling Street, in Corbridge, named from that famous road, in turn named after Wacol's people, but it is a modern import.

However, there are Roman names recorded. High Rochester was *Bremenium*, and inscriptions there refer to the *exploratores Bremenienses*, or Bremenium scouts, who

5 Signpost on the Military Road, south of Hadrian's Wall

patrolled the area north of the Wall; Rudchester was *Vindobala*; Halton Chesters, *Onnum*; Chesters, *Cilurnam*; Carrawburgh, *Brocolitia*; Housesteads, *Vercovicum* Great Chesters, *Aesica*; Carvoran, *Magna*. Corbridge for long was known as Corstopitum, but more recent investigation shows it as *Coria*. It is through the Vindolanda writing tablets that more information on names will come to light, not only local, but national. The name Vindolanda is clearly authenticated in these vital documents.

Scholars are uncertain of the meaning of the surviving Roman names, but there is agreement on some of them.

From east to west, Rudchester is *Vindobala*, and it is possible that the first element means white (as in Vindolanda) or shining, and *bala* is a place. The 'white' may have to do with vegetation or the openness of the site as it caught the sun, but this is uncertain. Haltonchesters was *Onnum*, and may be named after a British word for water. A stream runs by. Hunnum is a personal surname still in use. Chesters was *Cilurnum*, and may mean a large expanse of water. This would make sense as the Tyne flows past, and had to be bridged. Carrawburgh was *Brocolitia*, a place with heather or a rocky place. The site already had the wall, ditch and vallum before the fort was built over the vallum.

Housesteads was *Vercovicum*. Built on a spectacular site over the whinstone ridge, the name may mean that it was a place of good fighters. A *vicus* in names today refers usually to the civilian settlements outside the forts, and the Old English -wick refers to a farm or settlement. Great Chesters was *Aesica*, one of the smallest forts, and its name may refer to Esus, a local god, but this is uncertain. If it did, there may have been a temple there.

Essential to the history of the Wall, but lying to the south of it on the old Antonine frontier are the large, important bases of Vindolanda and Corbridge. The name Vindolanda is actually derived from British: *vindos* means white. *Landa* means lawn or enclosure.

There is no natural feature that is white, so there might have been a type of white grass growing there. This is also a problem in Old English names which, like Whittle ('white hill') have the same element. Another possibility, put forward by Anthony Birley, is that as the lawn lay in shadows from surrounding hills, frost melted everywhere except on the grass, then shone in the sun. It's an attractive poetic explanation, but only one possibility. Corbridge was *Coria*, but the meaning is unknown. The Cor Burn takes its name from the site, and not vice versa.

Further afield, to the north is Risingham, called *Habitancum*, a name found on an altar, possibly a personal name like Habitus. High Rochester was *Bremenium*, meaning a roaring stream. To encroach on Cumbrian territory for a moment, Bewcastle is one name where there is no doubt about what it means. It was *Fanum Cocidii*, the temple of Cocidius, to whom altars were raised. That concludes the little known about Roman names, but it is possible that sources like the Vindolanda tablets will produce more in the future. Meanwhile, the Wall zone has many other names to enjoy and puzzle over, unconnected with the Romans. A famous one is 'Goodwife Hot', where the second element is nothing to do with her high temperature or inclinations, but means a little 'holt' or wood.

SETTLERS FROM NORTHERN EUROPE

With the internal collapse of the Roman Empire, what had been ruled by Rome became attractive to raiders, who later became settlers. The so-called 'Anglo-Saxons' and Jutes were, not so long ago, credited with the establishment of England's agricultural landscape, but this is not true. Before the Romans, much of today's landscape was already formed, and the Romans saw this as a good reason for invading, as the pickings were rich. A system of villas with their estates continued to exploit rich farmland, encouraged by a move away from the high taxation of towns. In turn, the new predators, sensing the weakness of Empire when the Roman troops withdrew, found a relatively undefended coastline of great length to pick on for their raids. Apart from pockets of local resistance and the efforts of the Romano-British leader who entered legend as 'King Arthur', the bear, the invaders were able to get their own way; then, sensing that it was better to settle the land and grow rich that way, they became colonisers.

Often spoken of as 'The Dark Ages', this transitional period was like many others, when not much is known in detail about what was happening. However, it was these events that led to the establishment of the farms, villages and towns and the names that we know today.

The impact of the language was immense in naming places, although the Germanic languages only gave us about 30 per cent of our words. The rest were to come through Norman French.

The tribes and adventure groups pushing into England from northern Europe spoke a Germanic tongue. We know nothing of what survived of the language of Roman Britain, and the fate of the 'British' is uncertain. Whether many were slaughtered, enslaved, or

6 Roman Wall west of Housesteads

were allowed to co-exist with the newcomers is not known. They themselves had tribal allegiances that were to cause conflict among them. Harold himself before marching south again to fight William of Normandy (a land originally settled by Norsemen) had to fight a battle against northern Europeans who were claiming his throne.

In the early fluid period of raids and settlement, family groups become identified through place names. No doubt the first places to be colonised were the easiest places to access, such as along the coast and along the river valleys. The most productive land was the first target, then, as more settlers came in and population increased, the marginal areas would have been settled. In the South, names such as Worthing used to be thought to be the oldest, identifying a person as a leader of his people. In Northumberland there is only one: Birling on the coast near Warkworth, but there are many 'ington' and 'ingham' elements, preceded by a personal name. These, again, are not necessarily the earliest, but the information that they carry – the names of people – is important. If it were not for this, there would be no record of settlers' names, at whatever time they came here. In the past, historians concentrated on names of this type, but there are thousands of others to be considered. These also identify features that they encountered: permanent ones such as hills, cliffs, rivers and streams. They tell us about conditions, such as boggy ground, light sandy soils, woodland, springs, vegetation, wild and domesticated animals. We are given a kind of Breughel painting of landscape and all that is in it, and these things change as agriculture changes, as population increases. There are both permanent and fluid features.

The most important thing for anyone studying names, whether as a specialist or someone just interested, is to go into the landscape and check the positions of the names, for how else can one relate them to physical features? If the supposed meaning does not

7 A wide selection of names signposted south of Hexham

fit the landscape feature, it has to be re-thought. For example, for the element *hoh* to be accepted as a ridge, there must be a ridge at that place, whether prominent or slight. A *burgh* can only exist where there has been a fortification.

A warning, though: spelling cannot be taken at its face value. We live at a time when spelling has become more or less fixed, but in the past it was very fluid, and a word can appear on the same map with different spellings. The earliest spellings of the word and the changes must be carefully considered. The English Place-Name Society has published many valuable county surveys that will help you and me at least to understand in part this fascinating aspect of our history.

PLACE NAME ELEMENTS

It is now time to look at these imported words, and to see what the elements tell us about settlement. Each word may give more than one piece of information; the principal elements are listed here:

PLACE NAMES AND FIELD NAMES OF NORTHUMBERLAND

Elements used in place names

The following list includes most of the elements, other than personal names, that occur in England. Examples are given of Northumberland names that contain these elements, and some examples demonstrate that it is not possible to guess from the modern spelling which elements are present; this information can only come from very early spellings.

The abbreviations used are OE (Old English) and ME (Middle English).

OE *ac* = oak; *acen, aecen* = of oaks. Acton

OE *aecer* = field, ploughed land. The Old Norse is *akr*. Linacres

OE *aeppel* = apple. Apperley

OE *aesc* = ash tree, a very common place name element in England. The Old Norse is *askr*; *aescen* = of ash. Ashington

OE *alor* = alder. Allerwash

OE *angr* = grassland. Angerton, Ingram

OE *baec* = back, and ridge. (Old High German *bah*). Backworth

OE *bean* = bean. This is fairly common. Beanley

OE *beg, beger* = berry. Barmoor

OE *beo* = bee. Bewick

OE *beorce* = birch. Birkenside

OE *beorg* = hill, mound (especially a burial mound). Old Scandinavian *berg, biarg* Wooperton

OE *bere* = barley, and corn. Berwick

OE *blaec* = black, dark-coloured (e.g. water, soil, the colour of a hill or forest). Blakehope

OE *boc* = beech. Bockenfield

OE *bolum* = tree trunks. Bolam

OE *boll, bold* = building. Shilbottle

OE *brad* = broad, wide. Bradford

OE *braer* = briar. Barrow Law, Brierdene

OE *bremel, braemel* = bramble.

OE *broc* = marsh, brook, stream. Usually brook in the North. Broxfield

OE *brocc* = badger. Brockley Hall

OE *brom* = broom. Broomley

OE *brycg* = bridge. Corbridge

OE *bucc* = buck deer. Buckton

OE *burg, burh* = fortified place. Burradon, Bamburgh

OE *burna (burne)* = spring, brook, stream. Burn is commonly used in Northumberland for a small stream. Brunton

OE *by* = settlement (Old Norse *byr, boer*). Byker

OE *byge* = bend of a river. Bywell

Old West Saxon *bygging* = building. Newbiggin

OE *byht* = bight, bend of a stream. Nesbitt

OE *byre* = shed, cattleshed.

OE *byrgen* = burial mound. Hebron, Hepburn

OE *caerse, cerse, cresse* = watercress. Cresswell

OE *cald, ceald* = cold. Coldmartin

OE *calf, cealf* = calf. Callaly, Callerton

OE *calfru, calfra* = calves, of calves.

OE *camp* = (from Latin *campus*, a field) an enclosed piece of land.

OE *carr* = rock. Carham, Carraw

OE *catt* = wild cat. It can also become a nickname. Catton

OE *ceaster, caester* = (from Latin *castra*) a city or walled town, usually a Roman fort. It can apply to prehistoric fortifications, as we see in the Chester Hill type of name. It also appears as -caster, -castle and -cester. Chesterhope

OE *celde* = (from *ceald* = cold. ME *kelde* = spring). It can become 'childefel'. Colwell

OE *ceole* = throat in the sense of gorge, valley,

neck of land. ?Chollerford
OE *ceorl* = freeman. Charlton
OE *cese, cies* = cheese. Cheswick
ME *chace* = chase, hunting ground.
OE *cietel* = kettle, and a deep valley. Chattlehope Burn
OE *cild* = child, young person. Chirland
OE *cipp* = beam, log. Chipchase
OE *cirice, cyrice* = church. Chirton
OE *cis, cisen* = gravel, gravelly.
OE *claefre, clafre* = clover. Clarewood
OE *claene* = clean. Clennell
OE *clif* = cliff, rock, steep drop, promontory, a slope, the bank of a river. Heckley
OE *cloh* = a ravine (dialect 'clough'). Catcleugh
OE *cnaepp* = hilltop, hillock.
OE *cnoll* = a knoll (small round hill).
OE *cocc* = cock, wild bird. Coquet, Cocklaw
OE *cofa* = cave, den. ?Coe Burn
OE *col* = coal, charcoal. Colpits
OE *copp* = hilltop. Hawkhope
OE *cot, cote* = cottage, shelter for animals (sheep). Hepscott
OE *cran* = crane. ?Cramlington
OE *crawe* = crow. Crawley
OE *croft* = enclosed land for tillage or pasture. Ancroft
OE *crok* = crook, bend (Old Norse *krokr*). Crookhouse
OE *cros* = cross. Meldon
OE *crumb* = crooked. Cronkley
OE *cweorn* = a mill, a quern. Wharmley
OE *cy* = cow. Kyloe
OE *cylen* = kiln. Kilham
OE *cyning* = king. Kenton
OE *dael* = valley. Dalton
OE *denn* = pasture. Pigdon
OE *denu* = valley, dene. Very common in Northumberland. Blagdon. Deanham
OE *deor* = deer. Heatherslaw
OE *dic* = ditch, moat, dike, embankment. Often describes a prehistoric rampart and ditch. Detchant
OE *dun* = down, hill, and sometimes hill pasture. Dunstan
OE *ea* = river (Old Scandinavian *a*). As a second element it usually occurs as -ey. Ewart
OE *ear* = gravel.
OE *earn* = eagle. Yarnspath Law
OE *eg, ieg* = island, including a dry island in a marsh, land in a stream or between streams. It can be confused with *ea* (river). Ponteland
OE *ele* = a small island. It appears as -eels and eales. Elyhaugh
OE *ellern* = elder tree. Elrington
OE *faer* = passage. ?Lindisfarne
OE *fag* = many-coloured. Falstone, Fawdon
OE *fealg, fealh, felh*. ME *falwe* = fallow, in the sense of ploughed land. Fallowlees
OE *fealu* = fallow, pale brown colour, yellowish. Fallowfield
OE *fearn* = fern. Fairnley
OE *feld* = open country, plain ; probably used in the sense of fields cleared from the old forests. It probably applies to an area larger than a leah. Felton
OE *fen* = fen, marsh. Fenton
OE *fin* = a heap of wood. Fenrother
ME *flat* = a piece of level ground. Flatworth
OE *fleot* = creek, inlet, estuary. Fleetham
OE *flode* = a channel, stream. ?Flodden
OE *fola* = foal. Fowberry
OE *ford* = ford. Very common. Ford
OE *ful* = foul, dirty, rotten. Philip
OE *gaec* = cuckoo (dialect gowk). Gowk Hill
OE *gara* = a gore, a triangular piece of land, a strip of land. Overgrass
OE *gat*; Old Scandinavian *geit* = goat. Yeavering
ME *garth* = yard, garden, paddock. In dialect *garth* can also mean a farm.
Old Norse, Old Swedish *gata* = a road
OE *geard* = fence, hedge, enclosure. Earle

OE *gos(a)* = goose. Gosforth, Goswick.

OE *grene* = green, a grassy spot. Grindon

OE *(ge)haeg* = hay, enclosed land. Hayning

OE *heath* = heath plants, uncultivated land, heather. Heddon.

OE *hafoc* = hawk. Hawkhill.

OE *haga* = fence, fenced enclosure. It can appear as haugh, hough, law.

OE *haga* = hawthorn. Hawden.

OE *halh, healh* = a corner, an angle, a secret place, recess, cave. In Northumberland it developed as haugh, which is a piece of flat alluvial land by the side of a river. Originally this would have applied to land so formed by the bend of a river. Beadnell, Bothal

OE *halig* = holy. Holystone

OE *hals, heals*, Old Norse *hals* = neck, used in many senses. The Old Norse meaning is a projection, a narrow piece of land. The general sense is of a promontory, headland, projecting piece of land.

OE *ham* = village, estate, manor, homestead. One of the most common second elements. Harnham

OE *ham(m), hom(m)* = a meadow, especially by a stream, an enclosed plot of ground – probably its earliest meaning. It is difficult to distinguish this from *ham* above. Fenham

OE *hamel* = bare, treeless, from 'maimed.' Humbleton Hill

OE *hangra* = a wood on the side of a steep hill. The old sense was slope. Hanging Leaves

OE *har* = grey, hoary. Harsondale

Old Scandinavian (ON) *haugr* = heap, mound (especially a grave mound), hill. It is not always easy to distinguish it from OE *hoh*.

OE *heafod* = headland, summit source of a stream, promontory. Haltwhistle

OE *heah* = high, tall. Capheaton

OE *heg, hieg, hig* = hay. Aydon

OE *hege* = hedge. ?Hazon

OE *helde, hi(e)lde* = slope. Learchild

ME *helm* = helmet, but used as a helmet-shaped cattle shed. Helm

OE *henn* = hen. Hinding Bum

OE *heope* = rosehip. Hepden Burn, Hepple

OE *heorot* = hart, stag. Hartburn

OE *here* = army, host (Old Scandinavian *herr* = the whole people). Harbottle

OE *hlaw, hlaew* = hill, mound, burial mound (with a personal name). The form appears as -low, -loe, -law. Sometimes it is -ley. Elilaw

OE *hliep, hlyp, hiep* = leap, a place to be jumped over, steep drop. Lipwood

OE *hlinc* = lynchet, a bank separating strips of ploughed land on a slope. Lynch Wood

OE *hlynn* = a torrent, and a pool (in ME *linn*). Lowlynn

OE *hoh* = heel, projecting ridge. In dialect this is *hoe, heugh* = a crag, cliff, precipice. In general its meaning varies from a slight rise to a steep ridge. Cambo, Dunsheugh

OE *holegn* = holly. Hulne Park

OE *hol(h)* = hollow, sunken, deep. Holburn

OE *holm* = small island, dry land in a fen. (Sometimes the spelling changes to ham.)

OE *hop* = Its usual meaning is the same as *hope* in dialect, a small enclosed valley, a branch valley, a blind valley. Harehope

OE *horn* = horn, corner, bend, tongue of land, horn-like projection. Horncliffe

OE *hraefn* = raven. Ramshaw, Ravensheugh

OE *hramsa* = wild garlic. Ramshope

OE *hreod* = reed. Redburn

OE *hroc* = rook. ?Rochester

Old Norse. *hross* = horse.

OE *hrucge* = woodcock. Rugley

OE *hrycg* = back, ridge (rigg). There are many 'riggs'.

OE *hus* = house. Newsham

OE *hwamm* = literally corner. A small valley or nook surrounded by high hills. Whitwham

OE *hwer* = kettle, cauldron. Chattelhope burn, Ketley Crag

OE *hwerfel, hwyrfel* = circle, whirlpool. It began as the fly-wheel of a spindle or a spiral. (Old Norse *hvirfill* means circle, crown of the head). Whorlton

OE *hwit* = white, light-coloured. Whitley

OE *hyll* = hill. Ogle

OE *hyrst* = Originally meaning brushwood, it has come to mean wood, a sandy hill or knoll, a wooded hill. Hirst.

OE *ing* = In the -*ingas* form prefixed by a personal name, it *associates* a place with a personal name. It may not belong to that named person (who may be dead), but is named after him.

Old Scandinavian *kelda* = spring.

OE *laecc, lecca* = stream flowing through boggy land. Cawledge Park

OE *laefer* = rush, yellow iris. Learmouth

Old Norse *land* = land, estate, property, district. It is also used specifically in describing the quality of soil or type of tenure, whether it is cultivated, and what grows there. Coupland

OE *lang* = long. Langley

ME *launde* = glade, pasture, lawn. Vindolanda

OE *leah* = originally an open place in a wood, a glade, it has taken on a meaning of open land that is used as arable, and another main sense as wood, forest. Common forms are -ley, -leigh, lea, lee, with plurals lees, leese, leazes, leam. Bradley, Cheveley

OE *leoht* = light, bright, light-coloured. Green Leighton.

OE *lin* = flax. Linacres

OE *mael* = mark, cross. Meldon

OE *(ge)maere* = boundary. Difficult to distinguish from *mere* (lake). Greymare Hill

OE *mearc* = boundary, boundary mark, border.

OE *meoluc* = milk (usually showing good pasture). Melkridge

OE *mere* = mere, lake. Boulmer

OE *mersc, merisc* = marsh. Owmers

OE *mid* = middle. A settlement between two others. Middleton

OE *micel* = large. Mickley

OE *mor* = moor, waste upland, fen. Morwick

OE *(ge)mot* = meeting place – of streams or people (moot).

OE *mutha* = river mouth. Alnmouth

OE *mylen* = mill. Melkington, Milbourne

Old Norse *myrr* = mire, bog. Charlton Mires

OE *naess* = headland, cape, ridge. Fawns

OE *neotherra* = lower.

OE *ofer* = river bank, border, margin, sometimes a steep slope or a ridge. Wooler

OE *ora* = border, margin (very much like *ofer*).

OE *paeth* = path. This is still commonly used. Morpeth

OE *pol* = pool, deep place in a river, tidal stream. The Welsh *poll* is the source of pow, a slow-moving stream. Powburn

OE *port*: from Latin *portus* = harbour, town. Portgate from Latin *porta* = gate.

ME *quarrere* = quarry. ?Whlrleyshaws

OE *ra* = roe-deer. Raylees?

OE *raew, raw* = row (of houses) hedgerow. Catraw

OE *read* = red. Rede River

OE *rima* = rim, border, bank, coast. Rimside Moor

OE *rith, rithe, rithig* = a small stream. (Old Low German *ritha, rithe*). Ritton

OE *rocc* = rock. Rock

OE *rod* = rood cross. Rodestane

OE *roth, rothu* = clearing. Roddam

OE *ruh* = rough, uncultivated ground. Rowhope

OE *ryding* = clearing, cleared land. Riding Mill.

OE *ryge* = rye. Ryal

OE *sae* = lake, sea. Monkseaton

OE *saete* = seat. Simonside

Old Norse. *saet* = a shieling, hill pasture.

OE *sath, sealh* = sallow (plant). Possibly Saltwick, or see below.

OE *salt, sealt* = salt (*saltere* = salt-worker, salt-seller). Saltwick.

OE *sand* = sand. Sandhoe

OE *scaga* = shaw, thicket, grove. Elishaw

OE *scald, sceald* = shallow. Shadfen,

OE *sceaft* = shaft, pole. Shaftoe

OE *sceap, scep, scip* = sheep. Shipley

OE *scearp* = sharp, rugged, steep. Sharperton

OE *sceat* = a strip of land, usually overgrown, or even a park. Bebside

ME *schele* = (dialect: sheel, shiel) hut, shed. Agarshill, North Shields

OE *scora* = shore in the sense of steep rock, bank, hill. Shoresworth

OE *screawa* = shrewmouse. Scrainwood ? or below.

ME *schrewe* = shrew, but also rascal, villain. Scrainwood?

OE *scyf, scelf* = rock, crag, ledge. Shilvington

OE *scyttel, scyttels* = bar, bolt, arrow. Chesters (Scytlescester)

OE *slaed* = valley, dell. Weedslade

ME *slag* = slippery with mud, muddy. Slaggyford

OE *spic* = bacon. There is, however, another similar word with quite a different meaning. Dutch *spik* is a bridge made of treetrunks or brushwood. Low German has 'spike', a brushwood causeway. Wansbeck River?

OE *spring, spryng* = spring, well. But ME *spring* also means a plantation.

OE *staener* = stony ground. The Stanners

OE *stall, steall* = place, stable, stall. Whittonstall

OE *stan* = stone, stones. This can refer to special stones, such as monoliths, Roman mileposts, paved roads (Roman), unusual stones that are renowned for their shape, size or position, stones used for meeting places, etc. (Old Norse *stein* = stone). Featherstone, Stannington

OE *stede, styde* = site. Stede is often used to refer to farms, especially dairy farms. Newstead

OE *stig* = path. Styford

OE *stigol* = stile, steep ascent. The word *steel* came from this, and means a ridge, precipice, steep path up a ridge. Steel Rigg

OE *stoc* = monastery, cell, or just place, especially a place dependent on another. Stocksfield

OE *straet, stret* = street, Roman road, or other paved road.

OE *strod* = marshy land overgrown with brushwood. Broadstruthers Burn

ME *strother* = marsh, swamp. Coldstrother

OE *suth* = south. North Sunderland (Suthland)

OE *swaer, swar* = heavy. Swarland

OE *swin* = swine, wild boar. Swinhoe

OE *thing* = meeting place, court (Old Scandinavian – assembly, parliament). It appears as ting-, fing-, thing . Dingbell Hill

OE *thorn* = thorn bush. Thorngrafton

OE *thorp, throp* = farm, hamlet. Probably *throp* was a dependent farm. *Thorp* is especially common in Denmark and Sweden, and in England thorpes are common in Danish settlements. Throphill

OE *thyrel* = perforated, having a hole. Thirlwall

OE *thyrne* = thorn bush (Old Scandinavian *thyrnir*). Farnham, Caistron

ME *tod* = fox. Tod Hill

OE *tot* = lookout. Tosson

OE *treo(w)* = tree, usually a prominent or unusual one. Trewick

OE *trog* = a trough, then hollow or valley shaped like a trough. Trows

OE *tun* = originally a fence or enclosure,

but it soon became a homestead, village or town. On the whole names ending in -tun tend to be later than -ham. Its most general meaning is homestead or village, but it can also mean an outlying farm. Alwinton

OE *twisla* = fork of a river, land in the fork of a river. Twizel

OE *ule* = owl. Ulgham, Outchester

Old Scandinavian *ulfr* = wolf (OE *wulf*). Wooden

OE *unthances* = without leave (it becomes unthank) – a squatter's land. Unthank

Old Scandinavian. *vra* = corner, nook, remote place. Silly Wray

OE *waegn* = wain, wagon. Wansbeck

OE *(gel) waesc* = wash. Sheepwash

OE *waesse* = wet place, swamp. Allerwash

OE *waet, wet* = wet. Weetwood

OE *walh, wealh* = Briton (plural *walas, wealas*) – also serf.

OE *wall, weall* = wall, especially referring to prehistoric and Roman forts, and to Hadrian's Wall. Wallsend, Walwick

OE *weard* = watch. Warden

OE *wearg* = outlaw, criminal, felon, and the place where they were executed. Wreighill?

OE *weg* = way, road. Stannington

OE *well, wiell, waell* = well, spring, stream. Weldon

OE *weah* = holy place, heathen temple. Wooperton

OE *(ge) weorc* = work, fortification. Wark

OE *wer* = wier, dam.

OE *wic* = (Latin *vicus*) dwelling, dwelling-place, village, hamlet, town, farm. Probably the most common meaning is dairy-farm. Alnwick

OE *winn* = meadow, pasture. Heddon (East and West)

OE *withig* = willow (there is a side-form OE *widig*, -widdy, -a withy). Weedslade

OE *worth* = originally a fence or enclosure, then homestead. It continued for a long time in forming place names. Warkworth

OE *wrecca* = outlaw. Ratchwood

OE *wudu* = wood, forest, timber. Coquet

8 Landscape: Humbleton Hill to the North Sea across the Milfield Plain

When each name is broken up into its elements, these tell us:

- Physical characteristics of the landscape
- Rivers and streams
- Vegetation, natural and planted
- Animals and birds
- People
- Dwellings and other buildings
- Colour, size and French names
- Historical information

There is going to be an overlap within these categories, but in general the system should be satisfactory.

Following this, there is an alphabetical list of Northumberland place names with their possible meanings. I have included the Tyne-Wear region, as this was originally part of Northumberland, for the purposes of place name study.

All the information is gleaned from the work of place name scholars, such as Professor Allen Mawer in the 1920s, and Eilert Ekwall. Place names in the landscape have been more recently examined by Margaret Gelling in 1984. Even more recently is the landmark publication of the Cambridge Dictionary of English Place Names, edited by Victor Watts, published in 2004, which has shed new light on many names. As more work is undertaken by scholars, it is likely that some of these meanings will be challenged and new interpretations given. That is the function of academic research.

Most personal names listed here are no longer in use. People get their names in all sorts of ways: from their parents, usually, but there may be a physical characteristic which is recorded, such as the colour of hair. A nickname may begin with some physical abnormality or from some unusual form of behaviour. Much later in Border history, people would take the names of their leaders, so we find lots of Armstrongs, Halls, Croziers, Charltons. These names defined a loyalty, or the people may have had no alternative than to ally themselves with the most powerful individuals, in self-defence. Tribalism is strong. We can to some extent understand these names: strong of arm, a cross-bearer, free peasant and hall-dweller, but when we are faced with the Angles, there is a lack of anything against which to set their names unless we are familiar with the intimate details of their homelands. Some whose names appear in Northumberland have no record elsewhere. If it were not for their names being preserved in place name records, we might not even know their names. They are locked in our maps.

Then there is the landscape which they raided and then settled. Either they would name what they found there or copy the name from the local people. Once a name of a new settler became well-established, subsequent dwellers might continue to use the original name, and *ingas* might well indicate this root.

Not much survived of the language of 'British' people, those who were here in Romano-British times, but the river names of England, Wales and Southern Scotland had a high survival rate in their earliest forms, some probably from deep inside prehistory. To these were added Old English and Old Norse names. Many settlements then took their names from rivers, which were crucial to transport and penetration of the land and full of fish. Springs fed burns, which in turn fed the rivers. The speed of rivers depended on their size, the height from which they fell and the volume of water that they contained. It was essential to know the safe crossing-places, by ford or bridge. Their banks had good alluvial soil on the flood plains, with trees such as willows, so important in basket-making. There are many different words that refer to water whether it be in rivers, springs, streams, ponds or lakes. Marsh was important too, for this attracted birds that could be killed and eaten. It might also be an obstacle or an opportunity, once it was drained, to grow food in its rich soil. Wetland management was important to early settlers.

All these things depend upon the general lie of the land. Hills were penetrated and flanked by valleys, and there are many different words for high places and valleys, the latter depending on their depth, width and what flanked them. They may be called deans, for example, or hopes, as we shall see. There are even more words for hills and for slight rises in the land that give wide views. There is a considerable amount of such undulating land in Northumberland as well as the more dramatic scarps.

Some landscape is often described as 'natural', but in fact it can be argued that everything has been affected and changed by people's activities over centuries. Even before arable agriculture was exploited, people living a nomadic life of gathering food, fishing and hunting, managed the land to suit them, by burning parts so that plants they particularly wanted were regenerated. In Northumberland the only possible 'natural' woodland allowed to self-generate is in stream valleys where the land is unsuitable for anything else. Ploughing, intensive use of the land, tree-felling, over-grazing and slight changes in climate have destroyed some previously fertile areas, leaving them as 'waste' or 'moor'. Some areas where there is little or no woodland may carry names that indicate that this has not always been so.

Agriculture has done most to change the landscape's cover. There are areas now abandoned as grain-producers and grassed over, where old field systems are very clear. Whereas the oldest may not be documented, medieval systems of rig and furrow forming the basis for a two- and three-field system of agriculture are not only still visible but remembered in field and place names, particularly in the former. There are words for clearings in woodland, for most arable land was tree-covered at one time.

Animals inhabited the land, and were hunted or herded. There were wild birds. Place names give many of these, particularly those which are most common and most useful. People built homes, extended settlements to new areas, built farm buildings and enclosures for stock, crops and animals, and these features may be named. I include temporary dwellings which go with pastoral activity – the 'shielings' or temporary dwellings of herdsmen and women, where they stayed close to their beasts in summer pastures, usually in the hills.

Finally, there are those names that tell us something of the history of the area, such as the presence of Roman and earlier forts, of church or monastic lands, of holy wells, of people's status. Some tell us of the presence of prehistoric burial mounds. The field names add considerable extra detail to all this.

PLACE NAMES AND FIELD NAMES OF NORTHUMBERLAND

There are a small number of French names either standing by themselves or added to 'English' names, showing the new order of things under the Normans. The Normans, however, seem content to have recorded the sounds of local names as near as possible in their own surveys. They reportedly massacred most of the population of Northumberland, so that the land lay derelict and devastated. Who then survived to inform them of the local names of places that had been destroyed? Details of these name categories follow. Only meanings are given, without the documentary evidence that leads to these meanings. Numbers in brackets refer to the number of places with the same name.

PHYSICAL CONDITIONS

Cambo: hill spur and crest
Cambois: curving bay
Capheaton: high
Carham: at the rocks
Carraw: rocky ridge
Chattlehope Burn: kettle-shaped valley
Chollerton: gorge
Clifton: hill slope
Coe Burn: rocky overhang, small cave
Cold Law: cold
Coldmartin: cold mere, pool
Coldwell (3): cold spring
Colwell: cold spring
Cragshield: crag
Crookdean: winding stream
Cronkley: crooked cliff
Crookham: bends of a river
Crookhouse: at the bends
Deanham: valley
Deanmoor: fen in a valley
Denton: valley
Denwick: valley
Dewley: dew falls heavily
Dews Green
Dipton: deep valley
Downham: at the hills
Druridge: dry ridge, dunes
Dryburn: stream that dries up
Dunstan: hill of stone
Eland: river island
Eshottheugh: ridge, spur

Fenton: fen
Fenham (2): fen
Fenrother: fen
Fenwick (2): fen
Flatworth: flat ford, shallows
Grottington: sandy hill
Haltwhistle: hill above fork in river
Hanging Leaves: a slope
Hangwell Law: overhanging
Harnham: rocky, stony
Harsondale: grey stone
Haughton: low lying river land
Hawick: high
Healey (3): high
Heaton (2): high
Heckley: high cliff
Helm: helmet-shaped
Higham Dikes: high
High Laws: high hills
Hoppen: at the enclosures
Horncliffe: cliff in a tongue of land
Horton (3): muddy
Houghton (2): on a spur
Houtey: hollow clearing
Howden: deep valley
Howick: high
Humbleton: bare-topped
Kirkley: hill
Knaresdale: rock
Langhope: long valley
Lipwood: steep slope, lip-shaped projection

Lucker: marsh
Mickley: large clearing (*micela*)
Middleburn: middle
Middledean: middle
Mindrum: ridge
Mitford: junction of streams
Moralee: swampy clearing
Morpeth: moor (or murder) path
Morwick: fen
Mosswood: moss
Nesbit (2): noselike bend
Ord: point of land
Overgrass: on the edge, margin
Philip: Foul (*fulhope*)
Plenmeller: bare hill
Pont: valley (Welsh)
Ponteland: island in a river or marsh
Rimside: on the rim
Rock: rock
Roddam: at the clearings
Ros Castle: moor (Welsh *rhos* or Irish *ros*)
Roughley: rough clearing
Rowhope: rough valley
Sandhoe: sand ridge
Sandyford: sandy ford
Seaton: sea
North Seaton: sea
Seaton Delaval: sea
Shadfen: in a hollow (shallow)
Shaftoe: shaft-shaped ridge
Sharperton: steep hill
Shilvington: rock, crag, ledge
Shoresworth: steep slope
Slaggyford: slippery with mud
Slaley: muddy
Smales: narrow
Snabdaugh: snab, little hill?
Snitter: cold, exposed, snowy
Snook Bank: ridge
Soppit: marshy path
Spindlestone: spindle-shaped pillar
Spithope: spit-shaped
Stamford: stony

Stamfordham: stony
Steel: steep
Styford: path
Swarland: heavy
Sweethope: sweet, pleasant valley
Throphill: farm hill
Trows: depressions? troughs?
Waskerly: marsh clearing
Weetwood: wet
Wharlton: hill farm
Wheatridge: white
Whirleyshaws: quarry
Whorlton: round hill
Windyhaugh: windy hedge or haugh
Wingates: pass where the wind blows
Wydon: wide valley

RIVERS AND STREAMS

River Alwin: pre-Roman
River Aln: to flow, to stream
Allendale: like Alwin
Birtwell: bright, clear spring
River Blyth: pleasant, gentle, merry
Bowmont Water: with bends
Brunton: stream settlement
Bywell: spring at the river bend
River Breamish: pre-Roman, to roar?
River Derwent: pre-Roman from 'oak'
River Coquet: where there are wild game birds, but originally red in colour
Devil's Water: black stream
Easington: River Yese
Erringburn: British, silver, bright
Ewart: enclosure on a river
Fleetham: estuary
Flodden: stream from a hill
Font River: pre-Roman
Ford: fording place
River Glen: clean, holy
Holborn: deep stream by a spur
Irthing river: pre-Roman

Kielder: violent water
River Low: with tidal pools
River Lyne: pour, flow, glide
Maggleburn: pre-Roman
Nanny River: pre-Roman
Ouse Burn: gushing
Pont river: from *pant* in a valley
Ray: possibly at the river
Rede River: the red one
Ritton: small stream (*rith*)
Seghill: haugh on the Sige stream
Sleekburn: muddy stream

River Till: to dissolve, flow
River Tweed: perhaps 'powerful'
Twizel: fork in the river, junction
River Tyne; 'dissolve', 'flow'
River Wansbeck: a causeway made of brushwood
Waren Burn (Pharned): with alders
Weldon: spring valley
Welton: twisting valley
Wooler: spring promontory, margin

Those names listed above form a large and important picture of the physical conditions encountered and modified by early settlers. The highest hills of Northumberland lie in the north-west in the Cheviot range. 'Cheviot' itself, obviously important as an outstanding landmark above the others, has not left us with any documentary evidence to tell us what it means, although it is early, and British. Many of the individual hills in the volcanic range have their own names, some of which are not very old. Humbleton, and similarly-named hills in England, is 'bare-headed', so we may assume that it was unencumbered by vegetation. This makes sense because it was the site of an early Bronze Age cairn and a number of strong Iron Age enclosures that took advantage of a steep natural ravine on one side. In 1170 it was *Hameldun*. Another hilltop, Skirl Naked, with no such early spelling, is likely to come from Old Scandinavian *skirr*, meaning bright or clear, and again this fits a hill with no vegetation cover which shines in the sunlight. Elsewhere, Clennell is 'clean' and, away from the Cheviots, Plenmeller is a bare hill.

There are many different words to describe hills and ridges. One distinctly northern word is 'Law' (*hlaw*) used both in the sense of hill and of burial mound, with numerous examples. In the South, burial mound is the usual meaning, but in Hangwell Law, Gled Law and High Laws, to name a few, these are all natural hills, and some have given their names to settlements. The village of Doddington lies below the hill called Dod Law.

Other names that refer to hills are Barley Hill, Brenkley, High Law, Kearsley, Throckley and White Hall (Cramlington). Wreighill is gallow's hill.

A 'dun' is another common name for a hill, and was probably used in early settlement names, but went on being used long afterwards, as there are so many of them in England. It is combined with other elements. Dunstan is a stone hill. Downham simply means 'at the hills'. It can appear as 'don' and 'den', but is not to be confused with 'valley'. In Warden, it is a lookout hill, despite the modern 'den' suffix, and the village below it with its ancient church takes its name from that hill. A dun's height can vary considerably, and duns are widely spaced out over the county. In Fawdon and Fallodon the hill is fallow- or many-coloured.

Rises in the ground may be minimal, but they are marked in names. Snapdaugh is a little hill or 'snab' in hilly country. Sharperton has a sharp edge. There are many ridges or

9 The River Till at Weetwood Bridge

spurs with the suffix -hoh, a piece of ground that projects sharply, such as Ingoe, Shaftoe, Sandhoe, and Prudhoe, all with information about landscape such as sandy and shaft-shaped. Prudhoe may refer to the man who settled there: Belsay and Duddo certainly did. There is another lovely name in the original of Snook Bank, which means the ridge on which a shackle-yard stood: *Schakelzerdesnoke* in 1264. Eshottheugh is a ridge with ash trees. A final example is Cambo, which includes both the ridge and the fact that it is camber-shaped. This is where Capability Brown, the great landscape gardener, went to school.

Mindrum is a ridge. Druridge is a dry ridge because it is a line of sand dunes. High places sometimes have the prefix -hea, high, as in Healey, Heaton, and Howick. There are rocky cliffs and ledges: Carraw, Knaresdale, Horncliffe, Shitlington, and other names where 'heugh' appears. Hangwell and Coe are overhangs or rock shelters.

The shapes of physical features are well represented in names like Cambois (pronounced cammus), meaning a curving bay; bends in streams occur at Crookham and Cronkley. Nesbitt is nose-shaped, Helm is shaped like a helmet, and Spindlestone is spindle-shaped.

Into this landscape of hills, ridges, slight rises and valleys, we can now pour water. When we come to look at names that are concerned with water, there are many to consider. *Waesse* (OE) is a wet place or swamp, as at Allerwash, where alders grew. It can carry the meaning of a place to wash sheep or clothes. *Eg* is an island in marsh. One element in Ponteland indicated this, an island in the River Pont, perhaps where it was marshy. Fens abound, as in Fenham, Fenwick, Matfen, Mousen and Pressen. There are many references to the fens that provided the stuff of mystery in 'Beowulf', where dwelt the enraged God-hating duo of Grendel and his mother. Much land since its early settlement had to be drained, and this provided rich soil in many cases. However, far from being a barrier to progress, marsh land also provided a bonus if the form of creatures that could be hunted, trapped and eaten. Settlements were set up on the edges of fenland: Morwick and Moralee, for example. Weetwood is the wet wood, Slaley and Swarland are heavy to plough, perhaps because of wet clay soil. Slaggyford was heavy and soggy, and Philip was foul.

Mor (OE) can either be barren upland or marsh, probably the latter originally. Morwick, Murton and Barmoor are examples. *Halh, healh* (OE) is common in Northumberland, usually in the dialect form of 'Haugh' (pronounced 'hoff or 'harf'). Although its general meaning is a corner or nook, here it is closely related to alluvial land on the sides of a river. Sometimes these flat lands were formed in the bends of rivers. It is related to *holh*, a hollow place. In the North it is associated with water, so the Tyne valley haughs make it clear that what is referred to is the large expanses of flat land that separate Hexham from the River Tyne (thus Haugh Lane) or those at Warden. *Halh* is common in place names and field names in England, and its meaning varies with its location, such as dry land in a valley or in a marsh. These are some examples in Northumberland: Etal (Eata's or grazing pasture), Humshaugh (Hun's), Hepple (rose hip), Bothal (Bota's), Broomhaugh, Greenhaugh, Henshaw (Hod's), Kirkhaugh (church), Barhaugh (barley), Howtell (wooded), Beadnell (Beda's), Tughall (Tugga's) and Seghill (Sige's). Pauperhaugh is an example on the Coquet (Papworth's).

Valley names are numerous. *Hop* (OE) means somewhere remote, enclosed, but in Northumberland it is used mostly to name valleys – not its main use elsewhere. 'Hope' is generally a blind valley: Milkhope is an example. Langhope is long and Chattelhope is kettle-shaped, the latter giving a possible meaning to the place where one of the finest pieces of prehistoric rock art in the world was found at Ketley Crag, Chatton. Harehope is the hare valley, which fits its position admirably, very much enclosed and secret. Most of the *hop* names are close to the Border, and spill into Scotland. There are Broomhaugh, Sweethope, Chesterhope, Emblehope, Ramshope (wild garlic), for example. Another OE word for valley is *slaed*, not common in Northumberland but represented in Weedslade (willow). *Denu* (OE) also refers to valleys, such as Wyden. It becomes dene or dean.

Water was life and rivers provided food, a highway for exploration and access, and their names endured even though some of their meanings became obscure. Many were taken over by the Anglo-Saxons from the British, and in some cases the elements -ham, -ton and -wick were added. There were many different words for watercourses; there may have been too much, too little, emphasis on where it came from, where it went, how it flowed and how deep it was – all things to be taken into account by good farm management.

A standard name for river was OE *ea* (ON *a*), referring to water that was bigger than a brook or stream. Hexham was established at *Hagustaldes ea ham*, the river being the Tyne, of British origin, meaning something like 'dissolve' or 'flow' – the same as in River Till. Other major rivers such as the Aln and Alwin (and the Allen) have equally old recorded names, but we cannot be sure what they mean.

Blyth is pleasant, Breamish may have roared and Kielder was violent; the Derwent is named after oak trees, the Glen is clear water and the Tweed is powerful. Other pre-Roman names hiding their meanings are the Maggleburn, the Nanny River and the Irthing. We know what Coquet means: the cocks' wood, presumably after the game birds that were attracted there, but an earlier form means that it was a red colour. The River Reed is red, from iron in the soil. The Wansbeck, another major river, may be named from a causeway made of brushwood. Among streams are the Devil's Water, which is black, while the Erringburn is silver. Holborn is deep, Sleekburn muddy, the Warren has alders, Ritton is small.

Streams flow from springs, the latter being recorded as 'wells', more common in settlement names than 'stream'. Weldon is a fine example, as it also shows how imagination takes over; there is a story of Scots raiders being unable to find Brinkburn Priory in fog, but when the monks joyfully rang their bells of deliverance, the Scots returned, sacked the priory, returned to their leader who thanked them with 'well done, lads'. It is, of course, the place where a spring empties its water into the River Coquet. Other spring names are Birtwell (bright), Colwell and Coldwell (cool), and Bywell (spring at the bend of the river). Some wells are reputed to be holy, and bear that name. Hawkwell is either named from a person or hawk, Tranwell from cranes and Cresswell from cress. The element *celde* for spring is rare and not to be confused with *helde*, which means a slope, as in Learchild, or Akeld. The Old Norse equivalent of celde is *kelda*, as in Threkeld, in Cumbria, and is found only in the north and Scotland.

A name for a stream is 'beck', but this is rare in Northumberland, although common in Cumbria. *Burna* (OE) is more usual, with many elements describing its surroundings; Woodburn is wooded, we have seen that Sleekburn was muddy and Swinburn includes 'pigs'. Personal names are commonly attached, as in Simonburn.

The joining of streams is evident in Mitford, where *gemyth* (OE), its first element, means a junction. Twizell has a similar meaning, but with a different root, *twisla* (OE), meaning a fork, a meeting place.

Other names for watercourses include *fleot*, *fleote*, referring to an estuary, inlet or small stream. Fleetham is on a small stream that drains into the sea at Beadnell. Old English *mere* is a pond, pool or lake. Land-locked water was most likely to be a pool. The problem with 'mere' is that it is difficult to distinguish it from *gemaer* (OE), which is a boundary (as in Greymare), and *mere* as a mare. One very important place is Coldmartin (*mere-tun*), where in prehistoric times the source of water was essential to upland grazing in the summer. Two small lakes lie in a marshy hollow surrounded by an outcrop of Fell Sandstone and thin soil. It may have also functioned as a boundary. Boulmer, on the coast, was the bull's mere.

Old English *pyll*, *pull* and *pol* may have been Welsh originally. At Powburn it seems to apply to a slow-moving stream, part of the River Breamish, that forms pools in the gravel deposits, and has become a major gravel quarry area.

Hamm (OE) is land hemmed in by water or marsh, such as a water-meadow formed in a river bend. *Holm* (OE) is raised ground in a marsh, a river-meadow or island. There is sometimes an interchange with *hamm*. Basically it is firm ground, and to go south for a moment the classic example is Durham (*Dunholm* in 1000) on a spectacular promontory site, but most holms are on low-lying land.

There are names for minor watercourses that appear especially in field names. 'Letch' is common – a slow-moving water channel that may be dug for drainage, originating in the Old English *laecc, laecce* – a stream or bog. A *sic* (sike) is a small stream and appears in field names too.

Water and marsh are frequently-occurring features of the landscape and are dominant. Not surprisingly, fords, bridges and causeways are the means of crossing them. There are many fords, which would have come before bridges (OE *brycg*). Corbridge's Roman and medieval bridges are close, but at different places along the Tyne – the former having recently been excavated, reflecting too the changes in the course of the River Tyne. The settlement of Coria served by the Roman bridge shifted to the east in Anglian times.

A causeway was also a means of crossing water or marsh. Causey Park is an example. The River Wansbeck suggests that a brushwood causeway was constructed to assist the passage of wains or wagons. The term 'ford' is seldom used on its own, though it was on the Till. Mostly it is used as a second element, as at Bradford (broad), Stamford (stone), Styford (the road that crossed it), Barrasford (by a grove), Hartford and Gosforth (geese). Warenford is rare because it names the river to be crossed.

VEGETATION AND CROPS

Acomb: oak
Akeld: oak
Akenside: oak
Allerhope: alder
Allerwash: alder
Allery Burn: alder
Aller Hope: alder
Angerton: grassland
Apperley: apple
Ashington: ash
Aydon: hay
Bardon Mill: barley
Barhaugh: barley
Barley Moor
Barmoor: berries (cranberry)
Barrasford: a grove
Barrow Law: briar
Beanley: beans
Benridge: beans
Berwick Hill: barley
Berwick-on-Tweed: barley

Birchope: birch
Birkenside: birch
Bitchfield: beech
Bockenfield: beech
Bolam: tree trunks
Brandon: broom
Branton: broom
Brierdean: briar
Broomhaugh: broom
Broomley
Caistron: thorn-bush
Clarewood: clover
Clennell: free of weeds (clean)
Coldstrother: marsh overgrown with brushwood
Cresswell: water cress
Eachwick: oats (or Acca)
Elrington: elder
Embley: elm
Eshells: ash tree
Eshott: ash grove

Espley: aspen
Fairnley: fern
Fallodon: fallow or newly-planted
Fallowfield
Fallowlees
Farney cleugh: ferny
Farnham: thorn bushes
Fenrother: heap of wood
Gatherwick: dogwood tree
Gorfen Letch: gorse
Grindon: green hill
Haughstrother: brushwood
Haydon Bridge: hay (or enclosure)
Hazelrigg: hazel
Heatherwick: heather
Heddon on the Wall: heather
Black Heddon: heather
Hedley on the Hill: heather
Hepden Burn: rose hips
Hepple: rose hips
Hetton: heather
Hirst: wood
Howtell: wooded
Hulne: holly
Humbleton: bare
Ilderton: elder
Ingram: grassland
Langley: long wood or clearing
Leam: groves
Learmouth: (River Lever, rush or iris)
Leighton: light
Lemmington: brook-lime (speedwell)
Linacres: flax
Longhirst: wood
Longwitton: wood
Lyham: grove-meadow
Lynch Wood: wooded ridge
Moralhirst: wood
Netherwitton: wood
Oakhaugh: oak
Oakwood
Ramshope: wild garlic (*hramsa*)
Redburn: reed
Redpeth: reed (or red)
Roseden: rush (*rysc*)
Ryal: rye hill
Ryle: rye hill
Shawdon: copse
Stobswood: tree stumps
Thornborough: thorns
Thorngrafton: thorns
Thornton: thorns
Thornhaugh: thorns
Trewitt: dry, resinous wood
Trewick: tree
Weedslade: withies
Witton: wood
Woodburn: wood
Woodhorn: wood
Yarridge: yarrow grass

When the Anglo-Saxon invaders arrived, they were not coming into a pristine wilderness, but to a well-developed agricultural landscape. The kinds of crops and areas of waste, fenland and moorland were self-evident, and a preponderance of a particular woodland, the amount of good grassland and other features would have made it possible for them to fell trees for houses and stockades, to feed domesticated animals, to allow them to hunt and fish as they had done in their homelands. The most valuable resource was arable farming, hand in hand with stock-raising. Crops of hay for winter feed were essential. The use of upland pastures, seasonally, was also essential.

The place names give us some idea of what conditions were like for natural growth and for crops. Forest and woodland were of crucial importance to the economy. Homes and farm buildings and some fences were timber-built. Wood was used for fires, for

10 Bockenfield House: the beech field

charcoal-burning, boat-building and for small bridges. The forest was also a food supply for gatherers, hunters and for animals such as pigs and deer.

Clearings in the forest or woods were important, for these became pasture and later were ploughed. It is a matter of debate how far Anglo-Saxon settlement was built on existing Romano-British sites (although this would seem the logical thing to do) and how far use was made of new clearances. Clearances were a means of providing new land for new wealth and for an increased population, so if the demand decreased (after the Black Death, for example), cleared land would become colonised by scrub and trees.

A grove (OE *bearu, fyrth, fyrthe*) was land overgrown with brushwood, known as a 'strother' locally. Strother is also a modern surname. Bolam, another surname, is the place where there were tree trunks, and Stobswood is tree stumps. *Gref, grafa, grafe* (OE) were also a grove or a copse, present in minor names, but common. Some elements show where woodland was situated. *Hangra* (OE) was a wood on a slope. *Hyrst* (OE) was a wooded hill. The first element in Howtel is *holt* (OE), a wood.

It is probable that individual trees captured attention when names were being given, as they could attract legends, mark places where things happened, or be associated with a particular person. But trees in general are named after their kind, showing us what was growing then. The most common were apple, oak, ash, alder, birch, elm, hazel, pear, hawthorn and willow.

Here are examples: Apperly (apple), Allerwash (alor, alder), Bockenfield (beech, *boc*), Hazelrigg (*haesel*), Thornborough (hawthorn), Weedslade (willow, *willig*). Another name for willow occurs in Sillywray, famous for 'The Last Horsemen', where the *vra* is Norse for a nook, or sheltered place, and the 'silly' is salix, present in some Cumbrian names.

Aespe (OE) is aspen, as in Espley, *boc* is beech, in Bockenfield, and *ellern* is elder in Elrington. One of the most important trees and bushes was the hawthorn, for its resilience and rapid growth. It was a good barrier in enclosures, and Hackwood was thus formed. The woods themselves were *widu*, *wudu* (OE); Witton confirms the earlier form, *widu*. The word occurs in Woodburn, Woodhorn, Weetwood, Lipwood, Harewood and Coquet (*cocwudu*).

In the wood there may have been an outlaw, as in Ratchwood (OE *wrecca*). Scrainwood may have housed either villains or shrews (OE *screanwena*). Encroachments into the woodland to provide more land for pasture and cultivation produced the *leah* (OE), which can mean forest, glade or clearing, the first step in its cultivation. Eventually these areas became known generally as fields, leazes being the plural of lea, and giving its name to one end of the Newcastle United football stadium as well as the name of the street where I live. At the end of a name, 'ley' signifies a field (ley-liners have adopted this term), with an indication of what was in it, such as: Lambley, Horsley and Callaly (calves). Clearings themselves became known as 'ridings' from OE *rodu*, as in Riding Mill. If a field is called by that name, it is most likely to refer to a clearance rather than to horse-riding, if it is old.

Small woods were *sceaga*, which became 'shaw' as in Shawdon and Ellishaw, but this must not be confused with a ridge. To add to the woodland, there are plants that grow wild, such as ferns, briars, broom, dogwood tree, gorse, heather, holly, speedwell, reeds, rushes and yarrow grass. Some plants would have had food value and others medicinal properties, such as watercress, wild garlic, cranberries and rosehips. Clover would make good grazing and help soil fertility. Among the cultivated crops listed are barley, beans, oats, rye and flax. Grassland, fallow and hay also figure.

ANIMALS, BIRDS, INSECTS

Beal: bee
Bewick: bee
Bickerton: beekeeper's farm
Brockley Hall: badger
Buckton: deer
Callaly: calves
Callerton: calves
Catcleugh: wild cat
Catraw: cat's row
Catton: wild cat hill
Cawledge: crow
Cocklaw: wild game bird
Cockle Park: wild bird
Cornhill: crane
Cowgate: cow pasture
Cramlington: crane
Cullercoats: doves

Cushat Law; woodpigeon
Embleton: caterpillar
Ewesley: blackbird
Fowberry: foal
Foxton: fox
Gosforth: goose
Goswick: goose
Harehope: hare
Harelaw: hare
Hartburn: hart, stag
Hartford
Harthope Burn
Hartley
Hartley Burn
Hartington
Hartside
Harwood House: hare

Harwood Shiel: hare
Hauxley: hawk
Hawkhill
Hawkwell
Heatherslaw: stag, deer
Heiferlaw: heifer
Herons Close: heron
Hinding Burn and Flat: hen (water fowl)
Horsley: horse
Kyloe: cow
Lambley: lambs
Longhorsley: horses
Lucker: sandpiper
Marden: mares?
Otterburn: otter
Ottercops
Outchester: owls
Owmers: owls
Ramshaw: raven
Ravensheugh
Raylees: deer
Rugley: woodcock (*hrucge*)
Sheepwash: sheep
Shipley: sheep

Shield Dykes (Swynleys): swine
Snipe House: swine pasture
Stot Fold: stud fold
Swinburn
Swinhoe
Todburn: fox
Todhill: fox
Todridge: fox
Tranwell: crane
Ulgham: owls
Ulwham: owls
Wooden: wolves
Yarnspath: eagle
Yearhaugh: fishery
Yeavering: wild goats

DAIRY PRODUCTS

Butter law
Cheeseburn: cheese
Cheswick: cheese
Melkington: milk
Melkridge: milk

Names included wild birds and animals: wildcat, wolves, fox, badger, otter, crane, crow, woodpigeon (cushat), owl, eagle, hawk, blackbird (ousel), sandpiper, heron, raven and woodcock. Wild goats give their name to Yeavering Bell, still there, as well as to Gateshead.

Hunted animals include hare and hart (stag, deer); game birds, too, were hunted. Bees could have been wild or kept. There were domestic animals: calves at Callaly and Callerton, sheep, a foal, horses, geese, swine, heifers and cows. At Cullercoats doves were kept in dove cotes or 'doocats' as local pronunciation would have it, as an extra source of meat in winter. Finally, the word caterpillar may be present in Embleton.

PEOPLE

Abberwick: Aluburg
Abshields: Abba
Acklington: Aeccela
Adderstone: Eadred

Aldworth: Ealda (or old)
Amble: Anna
Amerston: Eymundr
Anick: Egelwin

PLACE NAMES AND FIELD NAMES OF NORTHUMBERLAND

Ardley: Earda
Backworth: Bacca
Bamburgh: Bebba
Bavington: Babba
Beadnell: Beda
Bebside: Bibba
Bedlington: Bedla or Betla
Belford: Bella?
Belsay: Bill
Bingfield: Bynna
Birling: Byrla
Blenkinsopp: Blenkin
Bothal: Bota
Branxton: Branoc
Brenkley: Brynca
Brinkburn: Brynca (or edge)
Brotherwick: Brodor
Buston: Butel
Buteland: Bota
Cartington: Cretta
Catchburn: Caecca ?
Chatton: Ceatta
Cheveley: Ceofa or Cifa
Chevington: Ceofa or Cifa
Chibburn: Cilla?
Chillingham: Coefel
Chollerton: Ceola
Choppington; Ceabba
Coanwood: Collan
Coastley: Cocc
Corsenside: Crossan (Irish)
Cottingwood: Cotta
Cottonshope: Cotta
Dinnington: Dun (or on a hill)
Doddington: Dudda (or hill)
Dotland: Dot (Danish or Swedish)
Doxford: Doc
Duddo: Dudda
Dunsheugh: Dunn
Duns Moor: Dunn
Eachwick: Acca (or oak)
Earsdon (2): Eanred or Eored
Edderacres: Aethelred

Edington
Edlingham: Eadwulf
Eglingham: Ecgwulf
Elford: Eller (or eel)
Elilaw: Illa
Ellingham: Ella
Ellington: Ella
Ellishaw: Illa
Elsdon: Elli
Elswick: Aelfsige
Eltringham: Aelfthere
Elwick: Ella
Embleton: Aemele (or caterpillar)
Eslington: Esla
Etal: Eata (or grazing pasture)
Felkington: Feoluca?
Felton Hill: Fygla
Framlington: Framela
Gamelspath: Gamel (Scandinavian)
Garretlee: Gerard
Garret Shields: Gerard
Gilden Burn: Gildwine
Gofton: Gof
Gunnerton: Gunware (ON)
Hadston: Haeddi
Hauxley: Hafoc (or hawk)
Heddon E & W: Hidda
Hepscott: Hebbi
Hudspeth: Hud or Hod
Humshaugh: Hun
Ingoe: Ing
Isehaugh: Ina or Isa
Kearsley: Cyneher
Keenlyside: Cena
Keepwick: Kepe
Kidland: Cydda
Kimmerston: Cynemaer
Kirkwhelpington: Hwelp
Learchild: Leofric
Lesbury: physician/leech
Lilburn: Lilla
Lilswood: Lilla
Longframlington: Framela

42

Lysdon: Lida
March Burn: Merce?
Mason: Maerheard
Matfen: Mata
Molesdon: Moll or Mull
Monkridge: monk
Monkseaton: monk
Mousen: Mul
Nafferton: Nattfari (Norse)
Norham(orig. Ubbanford): Ubba
Nunnykirk: Nunna?
Nunwick: nun
Ogle: Ocga
Old and New Moor: Penda
Ouston(2): Ulfkell, Wulf
Ovingham: Ofa
Ovington: Ofa
Pandon: Pampi
Paston: Palloc
Pauperhaugh: Papworth
Pegswood: Pegg
Pigdon: Pica
Prendwick: Prenda
Prudhoe: Pruda(or 'proud')
Rennington: Regna
Riplington: Riplinga
Rochester: Hrofi (or *hroc*)
Rosebrough: Osburh
Rothbury: Hrotha
Rothley: Hrotha

Rudchester: Rudda
St John Lee: a bishop
SewingShields: Sigwine
Shilbottle: Shipley
Shilmore: Scufel?
Shittlington: Scyttel (or enclosure)
Shotton: Scot
Shotton-in-Glendale: hill of the Scots
Simonburn: Sigemund
Simonside: Sigemund
Swainston: Sveinne?
Tarset: Tir?
Thirston: Thrasfrith (pushy, thruster)
Throckley: Throca
Throckrington: Thoker
Thrunton: Thurwine?
Titlington: Tytel
Togston: Tocg
Tritlington: Tyrhtel
Tughall: Tugga
Wallington: Walh
Warkworth: Werce
Whittingham: Hwita
Whittington: Hwita
Widdrington: Widuhere
Wilkwood: Willoc
Willimontswick: Willimot (French)
Willington: Wifel
Woolsington: Wulfsige
Yetlington: Geatela

Anglo-Saxon names form a large part of place names. Without them, we might know nothing of many of those people listed above. Most appear as first elements, naming a settlement or landscape feature, and some names appear relatively frequently.

DWELLINGS/BUILDINGS

Ancroft: lonely croft
Bolton: with a building on it
Budle: a building
Coldcoats: cottage, sheep shelter
Earle: a yard, enclosure

Haining: an enclosure
Harbottle: building
Hoppen: at the enclosures
Lorbottle: with a building
Milton: mill

Newbiggin: new building
Newham: new farm
Newlands: newly acquired
Newsham: new houses
Newstead: new farmstead
Newminster: a monastery
Newton, Newtown: new settlement
Shelly: clearing with a shieling
Shield Hall: shieling
North Shields: shieling
Shieldfield: shieling
Snook Bank: shackle-yard on a ridge
Stanton: stone farm
Staward: stone enclosure
Stelling: animal enclosure
Trewick: built with wooden posts
Yardhope: yard, enclosure

The names above help to place people in homes, close to fields of animals and crops. The original field was *feld* (OE), as in Felton, and means large cleared spaces, or open country. Thus it would be treeless, without buildings, level in hilly country, free of marsh and used for pasture. In the second half of the tenth century this term was associated with large areas of cultivation that produced food by ploughing, sowing and harvesting. It was organised into strips and ploughed communally. It becomes associated as a name not just with open land but with ploughland. Its use as a first element is rare, and is mainly used as a final element after a description of its size, shape, surface, position in the land and colour. In Bitchfield there are birch trees, and in other names it can belong to someone or have wild creatures living there.

Where the element *land* (OE and Norse) is used, it is land – an estate or new arable. The later the term is used, the more casual its use. In field names it is used to name a strip in the open-field system, or just a piece of ground. Sometimes there is a specific reference to its use. Buteland, for example, is at the limit of cultivation, flanked by high moorland. Coupland meant that it was purchased (ON *kaupland*). If the landowner's name precedes it, if it is ancient, 'land' is most likely to be cultivated, as in Dotland. Sometimes the land is heavy to plough or wet, as in Swarland. Some of the earliest names describing the form of the land can be as early as those with personal names.

It is against this background of wild and agricultural land that we see some traces of places where people lived. Early houses would have been made of wood. The sight of crumbling Roman stone buildings with the remnants of heating systems must have been awesome to people who lived in wooden huts. We know from archaeology the structure of some of these, such as those with their bases dug into the earth, or the fine buildings of King Edwin's palace at Yeavering.

Most settlements remained as farms and did not develop much, but villages became the focal point for trade and meetings, as well as the Church later. There always had to be a focal point, provided in prehistory, for example, by henges, stone circles and hillforts. By late Saxon times the villages had open-fields with arable and mixed farming, and the more ancient farms must have remained scattered outside. In the North the two-field 'runrig' system was in operation, based on an infield nearest the village and an outfield of common pasture. In both fields there are strips and stints, the allocation of land for arable and pasture for the people living there. The three-field system was more usual, with one of the fields allowed to lie fallow so that it could regenerate. There was also watermeadow, woodland and waste shared by all.

These arrangements are best seen in the field name study in the second part of this book. Within this landscape a lonely croft was singled out, Ancroft, much later to become a plague-hit village.

Botl signifies stone buildings, in the case of Harbottle a castle built after the Norman conquest. One farm, Trewick, is built of wooden posts. New settlements and new buildings are marked and these appear also in field names. Yards and enclosures of different kinds – walled, fenced and hedged – are there, and a common element is the shieling, a name given to a temporary settlement for herdsmen during seasonal grazing.

Three groups of subjects now follow, not large, but significant:

COLOUR

Black Blakehope: black valley
Black haugh
Black Heddon: dark heather
Blagdon
Blake Law
Brownridge
Fallodon: fallow-coloured hill
Fallowfield: fallow-coloured open land
Falstone: yellow or multicoloured
Fawns: many-coloured
Greenhaugh: green
Glendue: dark valley
Whiteburn: white
Whitechapel
Whitehill
Whitchester
Whitfield: white
Whitely: white
Whittle: white
Whitwham: white
Vindolanda: white

FRENCH

Beaufront: beautiful brow
Beaumont: fine hill
Bellasis: beautiful site
Bellister: fine place
Blanchland: white glade (Normandy)
Bulbeck Common: Village of Bolbec
Carriteth: land used for charitable purposes
Causey Park: causeway
Darras Hall: from de Araynis
Gubeon: Gobyon family
Guyzance: Giunes, near Calais
Haggerston: prob. Old French Personal name
 from 'hagard', wild, strange.
Plessey: enclosed park
Pucherton: Puchardon, Normandy
Scremerston: Skirmer, 'escrimer', fencer
Seaton Delaval: De la Val , Normandy
Vauce: a Norman name

SIZE

Bradford: broad
Bradley
Broadstrothers
Lanton: long

Colours that are outstanding enough to cause comment appear in names which include black or dark places, perhaps with additional connotations, such as forbidding, frightening, applied to hills, valleys and dark heather. White, on the other hand, may indicate brightness, as we have seen at Vindolanda, and names including white may draw attention to some vegetation growth, such as cotton-grass or other grasses that turn white at the year's end in fields, hills and valleys. Blanchland, from French meaning 'white' refers to the colour of the land in Normandy, Blanchelande in 1165. Fallow or multicoloured is applied to some open lands, valleys and hills. References to the size of fields, settlements and fords are present, too.

In addition to Blanchland, other borrowings from French can be a first element, as the word 'beautiful' is applied to a feature. 'Causey' is a causeway, Plessey is an enclosed park. Some place names are imported with the families who gained land in England, as at Darras Hall, Haggerston, Seaton Delaval, Gubeon and Vauce.

FURTHER HISTORICAL DETAIL

Carrycoats: possibly fort in the wood
Charlton: free peasant
Chesterhope: fort
Chesters: fort
Chirland: belonging to a young nobleman
Colpits: coal pits
Corbridge: *Coria*
Coupland: bought land
Cowden: where charcoal was burnt
Coupen: basket for catching fish
Craster: fort inhabited by crows
Detchant: end of the ditch or wall
Dingbell Hill: fields of assembly (Old Norse)
Dinley: hill clearing
Dissington: on a ditch or moat
Ditchburn: burn by a ditch or wall
Espershields: burnt hut or pasture
Farne: from a distance
Featherstone: a cromlech (three uprights and capstone)
Felton: open fields
Flotterton: perhaps a raft bridge
Fotherly: sheep folder
Fourstones: standing stones
Glantlees: lookout hill
Glanton: lookout hill

Gloster Hill: bright place for a fort
Golden Pot: Golda? hollowed-out blocks of stone on boundary
Greymare: boundary mark
Hallington: holy
Halton: lookout hill
Harbottle: building for the hired men (army)
Harlow Hill: the mound of the people
Hebron: high burial mound
Hepburn: high burial mound
Hexham: see Introduction
Holy Island: Lindisfarne: settlers from Lindis
Holystone: holy stone (Benedictine abbey)
Holywell E and W: holy spring
Kenton: Royal manor
Kilham: at the kilns
Kirkhaugh: church
Kirkley: burial mound called Crick?
Kirknewton: church
Meldon: hill with cross or monument
Milbourne: mill
Newborough: new fort
Newcastle: castle
Portgate: gateway through the Roman Wall
Pressen: priest's fen
Preston: priest's farm

Prestwick: priest's farm
Reaveley: reeve (or rough)
Riding: clearing
Ridlees: cleared
Rochester: fort
Roddam: clearings
Rodstane: cross-stone
Rouchester: fort on rough ground
Rudchester: Rudda's fort (?)
Saltwick: salt
Scrainwood: villains or shrewmice
Settlingstones: where horses were mounted?
St John Lee: named after founder
Shortflatt: furlong
Stocksfield: belonging to a monastery
Sturton: paved road settlement (near Gloster Hill)
Sunderland: south land
Tecket: pre-Roman name
Tedcaster: fortification (Tada?)

Thirlwall: gap in the Roman Wall
Threepwood: disputed ownership
Thropton: farm at crossroads
Tone: on which a toll is paid
Tosson: lookout stone
Unthank (2): without leave, squatter
Walker: marsh near (Roman) wall
Wall
Walbottle: building on the wall
Walltown: settlement on the wall
Walwick: farm on the wall
Warden: watch, lookout hill
Wark (2): fort
Warton: lookout place
Wharmley: mill clearing
Whittonstall: quickset hedge
Wooperton: temple-hill valley
Wreighill: felon/gallows hill
Wylam: mechanical device, water mill

The final selection of names must include references to some aspects of local history, already in many cases a part of names already dealt with. Many ancient sites encountered by early settlers were named. Some were fortified and this accounts for the number of 'chesters', which applies either to Roman or pre-Roman sites and the element 'burgh'. Some were abandoned, as we see in those inhabited by crows.

Standing stones, which must have been visible at the time of the Anglo-Saxon settlement, were certainly added to names at a later date, but among the early ones are Fourstones, though no longer there. A Neolithic burial mound of the 'cromlech' type, with a burial chamber of three standing stones on which a big capstone was balanced, was constructed inside a long mound and seems to be the explanation for Featherstone. Other burial mounds still in the landscape, some of them excavated, belong to the early Bronze Age and still have the name Hepburn and Hebron to attest their height and holiness. In the case of Hepburn near Chillingham, a cluster of such mounds overlooks the site of others on Old Bewick Moor. Harlow Hill is the people's mound. Kirkley is Crick's mound or a hill.

There is a puzzle in the location of Gloster Hill, south of Warkworth, for, sharing its name with the more famous Gloucester in the South, no fort has been located there. However, a Roman altar came to light, so this may be a clue to a vanished site. Archaeologists must always be aware of such clues from the many sources at their disposal, of which place names are one.

Among special places in the landscape, some were meeting places; the rare Dingbell Hill is thought to be a place of assembly, a Norse term. Details of ownership including land belonging to the king, to priests, monasteries, noblemen, free peasants, a reeve, a named person, appear, as well as those that have attracted some sort of reputation for hiding outlaws and villains, or as places of execution.

Places may have served as lookout hills, prominent boundaries, or the end of a wall or ditch. They may be named after industries, such as lime-kilns, charcoal-burning, coal pits, salt and mills in addition to all those that we have seen marking fords and other crossing places of rivers and streams.

In Coupland we gather that the land was bought; elsewhere there is disputed ownership and the presence of squatters. As we have seen, there are indications of the large open-fields and their division into smaller units for farming such as furlongs. Clearings, 'ridings', are included, as are enclosures. Inevitably, there has been overlap in the examination of what some of the place names mean. When we turn to field names, there will be considerable reinforcement of these elements.

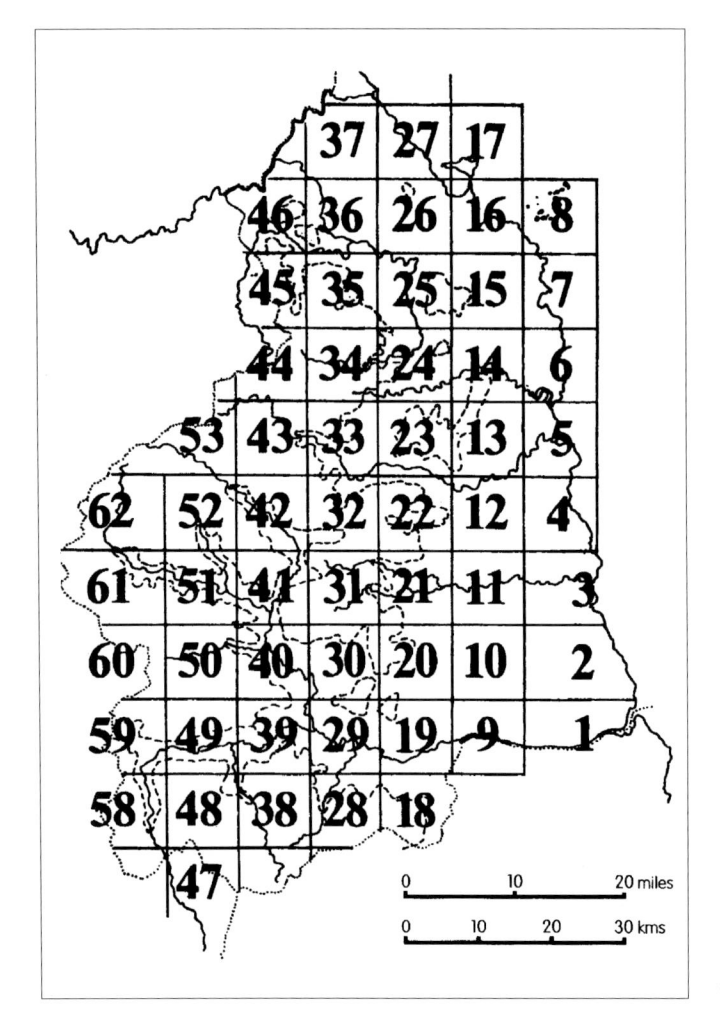

11 Map showing location of places on the grid

ALPHABETICAL LIST OF PLACE NAMES IN NORTHUMBERLAND

Abbreviations
OE = Old English, ON = Old Norse, OF = Old French, OS = Old Scandinavian, ME = Middle English.
* Names that have disappeared. ( ) Rivers located in more than one square area.

Sources
Ek = Ekwall, M = Mawer, W = Watts

Some elements, such as tun, wic and ham, are similar in meaning. All may have started as farms or 'settlements'.

14 Abberwick	Alburwick 1170	The *wic* or dwelling of Alu(h)burg (woman)
12 Abbshiels	Abscheles 1286	Abba's house
5 Acklington	Eclinton 1177	The settlement named after Aeccela
29 Acomb	Akum 1268	OE *acum* = at the oaks
28 Acton	Akedene 1269	OE *ac-denu* = oak valley
13 Acton	Aketon 1242	OE *ac-tun* = oak farm or Acca's tun
16 Adderstone	Edredeston 1233	Eadred's settlement
48 Agars Hill	Algerseles 1278	Ealdgar's house (M)
35 Akeld (ay-keld)	Achelda 1169	OE *ac-helde* = oak-slope (M thinks it might be a spring forming a marsh on the edge of the Till valley)
32 Akenside	Akenside 1332	The side of a hill covered with oaks
11 Aldworth	Aldewurth 1120	Ealda's or old *worth* (enclosure)
(38) Allen River	Alwent 1275	Like the River Alwin, to flow
(38) Allendale	Alwentedal 1226	Allen valley
38 Allenheads	Allenheads	Sources of the River Allen
37 Allerdean	Elredene 1099	Aelfhere's valley (M), alder valley
34 Allerhope Burn	Alrehopeburn 1240	Alder valley
39 Allerwash	Alrewes 1202	OE *alra-waesse* = alder swamp or alluvial land
(26) Allery Burn	Alriburn 1292	OE *alra-burn* = burn of the alders
(14) Aln	Alaunos 150, Alne 730	Pre-Roman: to flow, stream
34 Alnham	Alneham 1228	Settlement on the Aln
6 Alnmouth (almuth)	Alnemuth 1201	Mouth of the Aln
14 Alnwick (annik)	Alnawick 1160	Farm on the river Aln
(33) Alwin R	Alewent 1200	Pre-Roman name

PLACE NAMES AND FIELD NAMES OF NORTHUMBERLAND

33 Alwinton	Alwenton 1233	Settlement/farm on the river Aln
5 Amble	Ambell 1204	Anna's promontory
6 Amerston	Aymunderston 1243	Eymundr's farm (M)
27 Ancroft	Anacroft 1122, Anecroft 1195	Lonely croft
21 Angerton	Angerton 1186	OE *angr-tun* = grassland settlement
29 Anick (ay-nik)	Aeilnewic 1160	Egelwin's (Ek) or Aethelwine's farm
18 Apperley	Appletreleg 1201	Apple tree clearing
28 Ardley	Herdeley 1228	Earda's clearing (M)
3 Ashington	Essende 1170, Essenden 1205	OE *eascen-denu* = ash tree valley
19 Aydon	Ayden 1225, 1242	OE *heg-denn* = hay pasture
14 Aydon	Aydun 1279	OE *heg-dun* = hay hill
28 Aydon Shiels	Aldenschelels 1341	Eadwine's shieling (M)
58 Ayle Burn River	Alne 1347	Aln
2 Backworth	Buxwortha 1203	Bacca's enclosure
42 Bagraw	Bagraw 1385	A hawker's row of houses
16 Bamburgh	Bebbanburh 890	Bebbe's or Bebban's fortification. Bebba was wife of King Aethelfrith of Bernicia
48 Barhaugh	Berhalu 1279	OE *bere-halh* = barley haugh
18 Barley Hill	Birlawe 1255	Barley hill
26 Barmoor	Beiremor 1231	OE *beger* = cranberries
30 Barrasford	Barwisford 1231	OE *bearu, berwe* = by a grove
33 Barrow Law	Brerylawe 1304	ME *brere* OE *hlaw* = briar hill
24 Barton	Barton 1199	OE *bara-tun* = bare farm, or barley farm
31 Bavington	Parva Babington 1242	Babba's named settlement
7 Beadnell	Bedehal 1161	Beda's (Ek) or Bedwine's (M) *halh* (haugh)
27 Beal	Behil 1208	OE *beo-hyll* = bee hill
24 Beanley	Benelegam 1150	OE *bean-leah* = beanfield
29 Beaufront	Beaufront 1356	Beautiful brow (of the hill)
30 Beaumont	Beaumont 1232	French = fine hill
3 Bebside	Bibeshet 1198	Bibba's *sceat* or *(ge)set* = land or dwelling
3 Bedlington	Bedlington 1050	Settlement named after Bedla or Betla
10 Bellasis	Beleassis 1279	OF *assise* = beautiful site
16 Belford	Beleford 1242	Either Bella's ford or settlement at a bell-shaped hill
41 Bellingham (injm)	Bainlingham 1170	Hill-dwellers' farm (Ek) or Bel's
49 Bellister	Belester 1279	OF *bel-estre* = fine place (Ek)
43 Bell Shiel	Belleshope 1330	Bell's valley (M)

PLACE NAMES AND FIELD NAMES OF NORTHUMBERLAND

10 Belsay	Bilesho 1162	Bill's (Bilfrith's) *hoh* = ridge, spur
11 Benridge	Benerig 1242	OE *bean-hryg* = ridge where beans grew
1 Benton, Little, Long	Bentune 1190	OE *beonet-tun* = coarse grass, or bean farm
1 Benwell	Bynnewalle 1050	OE *bionnan-walle* = beside the Roman wall
27 Berrington	Berigdon 1208	OE *byrigdun* = fortified hill or berry hill
10 Berwick Hill	Berewic 1205	OE *bere-wic* = barley farm
27 Berwick-on-Tweed	Berewich 1167	Barley farm
25 Bewick	Beuuiche 1136, Bowich 1167	OE *beo-wic* = bee farm
33 Bickerton	Bikerton 1236	OE *beocer-tun* = beekeeper's farm
33 Biddlestone	Bitlesden 1181	OE *bytle/botl-dun* = valley dwelling
1 Billy Mill	Molendinum de Billing 1320	Billing's Mill (M), or by a ridge (W)
6 Bilton	Bylton 1242	Settlement at a hill edge
30 Bingfield	Bingefeld 1181	Open field named after Bynna, or by a hollow
15 Birchope	Byrchensop 1325	? Beorthtwine's or birch valley
18 Birkenside	Byrkinside 1262	Birch slope
5 Birling	Berlinga 1187	Byrla's people
20 Bitchfield	Bechefeud 1242	OE *bece-feld* = beech field, open land
52 Black Blakehope	Blachope 1230	OE *blac-hop* = black valley
*16 Blackmiddingmore	Blacmyddingmore 1333	ME *middying* = a dung heap
13 Black Lough	Blakemere 1200	Black mere
2 Blagdon	Blakedenn 1203	OE *blaec-denu* = black valley
25 Blakelaw	Blakelawe 1251	Black Hill
28 Blanchland	Blanchelande 1165	French = white glade (Normandy)
59 Blenkinsopp	Blencheneshopa 1178	Blenkin's valley or Welsh *blaen* = top
(3) Blyth River (blye)	Blitha 1204	OE *blithe* = gentle, merry, pleasant
3 Blyth	Blida 1130, Snoc de Bliemue	On the projecting land at the river mouth
12 Bockenfield	Bokenfeld 1206	OE *bocen-feld* = beech field
21 Bolam	Bolum 1155	OE *bolum* = the tree trunks, or *bol-ham* = rounded hill farm
14 Bolton	Bolton 1200	OE *bold-tun* = a place with a building on it
3 Bothal	Bothalle 1233	Bota's *halh* = land by the river
6 Boulmer (boomer)	Bulemer 1161	OE *bulan-mere* = bullock's or bull's pond
46 Bowmont Water	Bolbenda 1050	A bend in the river
37 Bowsden	Bolesden 1195	OE Boll's valley

16 Bradford	Bradeford 1212	Broad ford
20 Bradford	Bradeford 1242	Broad ford
49 Bradley	Bradeley 1279	Broad clearing
5 Brainshaugh	Bregesne 1104	From *borrans* = burial mound
24 Brandon	Bremdona 1150	OE *brom-dun* = broom hill
14 Branton	Bremetonam 1135	OE *bremen-tun* = broom farm
46 Branxton	Brankeston 1195	Branoc's settlement
(24) Breamish River	Bromic 1040	Pre-Roman, like Welsh *brefu* = to roar
2 Brenkley	Brinchelawa 1178	Brynca's *hlaw* (hill or mound) or 'brink'
4 Brierdene	Brereden 1295	Briar valley
12 Brinkburn	Brinkeburne 1188	Either Brynca's burn or the edge
35 Broadstrothers Burn (Cheviot)	Bradstoir 1255	OE *brad-strod* = broad strother
15 Brockley Hall	Brockleygehirst 1309	OE *brocc-hyrst* = badger wood
19 Broomhaugh	Brunhalwe 1242	Broom-covered haugh
18 Broomley	Bromley 1242	OE *brom-leah*, a broom grove/clearing
5 Brotherwick	Brotherwyc 1242	Brodor's farm
25 Brownridge	Brunrige 1330	OE brown ridge
7 Broxfield	Brokesfeud 1256	Field on the brook
7 Brunton	Burneton Batyll 1242	OE *burna-tun* = settlement by the burn
1 Brunton (E,W,N,S)	Burneton 1242	Brook farm
26 Buckton	Buketon 1208	Bucca's, deer, or goat farm
16 Budle	Bolda 1165, Bodle 1196	OE *botl* = dwelling, building
28 Bulbeck Common	Bolebec 1236	Named after the Norman village Bolbec, at the mouth of the Seine
2 Burradon(rdn)	Burgdon 1242	OE *burh-dun* = a hill with a fortification
33 Burradon	Burwedon 1242	OE *burh-don* = a hill with a fortification
16 Burton	Burton 1242	OE *burh-ton* = a fortified settlement
5 Buston, High-, Low-	Buttesdon 1166	OE *Buteles-dun* = Butel's hill?
41 Buteland	Boteland 1242	Bota's land
9 Butter Law	Buterlaw 1242	Butterhill
1 Byker	Bikere 1196	OS *by-kiarr* = the village marsh
19 Bywell-on-Tyne	Biguell 1104	OE *byge-wella* = spring at the bend of the river
33 Caistron	Cers 1160, Kersten 1184 (M)	ME *kers*, OE *thyrne* = thorn bush by the carse (marsh)
23 Callaly	Calualea 1161	OE *calfa-leah* = pasture for calves
10 Callerton	Caluerduna 1100	OE *calfra-dun* = hill where the calves grazed
9 Black Callerton	Calverdona 1212	OE Calves' hill

10 High Callerton	Calverdon 1242	OE Calves' hill
21 Cambo	Camho 1230	OE *camb-hoh* = hill spur with a crest
3 Cambois (kammus)	Cammes 1105, Cambus 1204	Welsh *cemmaes* = a bay, from Old Celtic *kambo* = crooked. Cambois stands on a curving bay.
21 Capheaton	Magna Heton 1242 Cappitheton 1454	OE *hea-tun* = settlement on high land Latin *caput* = chief (village)
44 Carham-on-Tweed	Carrum 1040	OE *carrum* = at the rocks (Ek) or homestead by the rock
40 Carraw	Charrau 1279	OE *carr-raw* = a rocky ridge
51 Carriteth	Le Caryte 1325	Old Norman French *carite* (*dh*) = land used for a charitable or religious purpose
47 Carr Shield	Carr Shield 1866 map	OE *carr* = shieling at the rocks
23 Cartington	Cretenden 1220	OE *cretta-dun* = the hill of Cretta's people
30 Carrycoats	Carricot 1245	? Celtic *caer-y-coed* = fort in the wood (M)
3 Catchburn	Cacheborn 1279	OE Caecca's burn?
31 Catcherside	Calcherside 1270	ME *caldchere-side* = cold-cheer hill
53 Catcleugh Shin	Cattechlow 1279	OE *catte-cloh* = the 'clough' or ravine of the wild cats. 'Shin' is Scots for a steep hill
2 Catraw	Catrawe 1479	OE *raw* = cat's row or Catta's row
38 Catton	Catteden 1225	Wild-cat valley
12 Causey Park	La Chauce 1242	ME *caucie, cause* from French *chausee* = causeway
4 Cawledge Park	Caweleg 1241, Cauleche 1252	ME *leche, lache* = long, narrow swamp through which water moves slowly. Crow.
41 Charlton	Carlton 1195	OE *ceorla-tun* = freeman's farm
15 Charlton, North-, South-	Charleton del North, Suth 1242	OE *ceorla-tun* = freeman's farm
53 Chattlehope Burn	Chetilhopp 1320	OE *cietel-hop* = kettle-shaped valley
25 Chatton	Chetton 1177	OE *ceatta-tun* = Ceatta's farm
20 Cheeseburn Grange	Cheseburgh 1286	OE *burg, burh* = fortification. Cheese
31 Chesterhope	Chestrehop 1298	OE *hop* = valley (by the fort)
30 Chesters	Scytlescester 1104	OE *scytles-ceaster* = fort used as an enclosure, or Scytel's fort
27 Cheswick (chizik)	Chesewic 1208	OE *ciese-wic* = cheese farm
Chew Green (Cheviot)		Narrow valley, cleft

4 Cheveley	Chiveleye 1300	Ceofa's or Cila's *leah* (glade, clearing, field)
4 Chevington	Cebbingtun 1050	OE Ceofa's or Cifa's named settlement
35 Cheviot (Chev,Chiv)	Chiuiet 1181	This is a pre-Roman name and its meaning is not clear
4 Chibburn	Chibrnemue 1228	?Cila's stream
25 Chillingham	Cheulingham 1186	OE Settlement named after Ceofel
40 Chipchase	Chipches 1229	OE *cipp-ceas* = a log structure. such as an animal trap. (Ek)
51 Chirdon	Chirden 1255	OE Burn (or bend) in a valley (W)
1 Chirton	Cherlton 1203	OE *cyrictun* = land belonging to a church
23 Chirland (Chirnells)	Childerlund 1178	OE *cildra-land*. *Cild* = child. It could be land belonging to a young noble awaiting knighthood (as in Childe Harold, Wynd)
30 Chollerton, -ford	Choluerton 1175	The ford is probably first named, the village belonging to Ceola, in a gorge
3 Choppington	Cebbington 1040	OE *Ceabbing(a)tun* = Ceabba's named settlement
19 Clarewood	Claverworth 1212	OE *claefre-worth* = clover enclosure
33 Clennell	Clenhil 1181	Literally 'clean hill'
3 Clifton	Clifton 1242	Farm on a hill or a slope
58 Coanwood	Collanwode 1279	OE Collan's wood
29 Coastley	Cotisley 1250	OE Cocc's clearing (M)
13 Cocklaw	Creklawe 1296	OE *cocc-hlaw* = wild bird hill
4 Cockle Park	Cockhill 1314	OE *cocc-hlaw* = wild bird hill
23 Coe Burn	Coveburn 1295	There are rock overhangs and small caves
10 Coldcoats	Caldecotes 1242	Cold cottages or sheep shelters
(33) Coldaw Burn	Caldelauburne 1255	OE cold stream
25 Coldmartin	Calemerton 1195	OE *mer-tun* = farm by the cold mere or pool
20 Coldstruther	Caldestrother 1232	OE cold, marshy ground overgrown with brushwood
21 Coldwell	Colewell 1277	Cool stream or spring
30 Coldwell	Caldewell 1325	Cool stream or spring
3 Coldwell	Caldewell 1242	Cool stream or spring
28 Colpits	Colpittes 1255	Coal pits
30 Colwell	Colewell 1236	OE *coh-wielle* = cool spring
(13) Coquet River (Coekt)	Coccuveda *c.*700, Coquedi Fluminis *c.*900, Cocwuda *c.*1040 (W)	The early meaning was British *cocco-wedd* = red river. It was changed to:

	Cocwudial 1050, Coqued 1104	OE *cocwudu* = woodland where cocks or game birds live
5 Coquet Island	Insular Coket 1135	Coquet Island
(23) Coquetdale	Coketdale 1160	The valley of the River Coquet
29 Corbridge	Corebricg 1050	The Roman settlement was 'Coria'
46 Cornhill-on-Tweed	Cornehale 1180	OE Crane's *halh* (haugh)
31 Corsenside	Crossinset 1254	? Irish personal name Crossan('s) hill pasture (ON *saetr*)
11 Cottingwood	Cotingwud 1257	OE Cotta's peoples' wood (M)
53 Cottonshope	Cotteneshopp 1230	OE Cot(t)en's or Cot(t)a's valley or field
36 Coupland	Coupland 1242	ON *kaupland* = bought land
30 Cowden	Colden 1250, 1286	Either OE *col-denu* = valley where charcoal was burnt or *col-denu* = cool valley
1 Cowgate	Cougate 1290	'Stint' of pasture for a cow
3 Cowpen (oo)	Cupum 1175	OE and ME *cupe* = basket for catching fish
40 Cragshiel	Le Cragscriel 1291	OE shiel by the crag (M)
2 Cramlington	Cramlingtuna 1130	OE *cranwella* = crane's spring ?
6 Craster	Craucestre 1242	OE *crawe-ceastre* = old fort inhabited by crows
24 Crawley	Crawelwe 1225	OE crow's hill
4 Cresswell	Kereswell 1234	OE spring where watercress grew
31 Crookdean	Crokeden 1324	OS *kroh*, OE *denu* = winding stream valley
18 Cronkley	Crombeclyve 1268	OE *crumbe-clif* = crooked cliff (M)
36 Crookham	Crucum 1244	OE settlement on the bends of the river
36 Crookhouse	Le croukes 1323	The bends (in the Beaumont Water)
2 Cullercoats	Culvercoats 1600	OE *culfre-cots* = dove cotes
34 Cushat Law	Cousthotelaw 1200	OE *cuscote* (dialect cushat), wood pigeon
28 Dalton	Dalton 1256	OE *dael-tun* = valley farm
10 Dalton	Dalton 1201	OE *dael-tun* = valley farm
10 Darras Hall	Calverdon Araynis 1242 Calverdon Darreyne 1346	A part of Callerton belonging to the de Araynis family from Airaines in Somme
42 Daveyshiel	Davisel 1344	OE Davy's shieling (M)
21 Deanham	Danum 1198	OE *denu-ham* = valley settlement
13 Deanmoor	Denemora 1280	OE *denu-mor* = fen in a valley
1 Denton	Dentuna 1252	Valley farm
6 Denwick	Denewyck 1242	OE *denu-wic* = valley farm
(18) Derwent River	Dyrwente 1040	Pre-Roman from British *derva* = oak
26 Detchant	Dichende 1166	OE *dic-ende* = end of ditch or wall

29 Devils Water (Tynedale)	Divelis 1233	British *dubo* = black; Old Welsh *gleis* = stream
9 Dewley	Deuelawa 1251	OE dew-hill
48 Dews Green	Dewegreane twelfth century	Dew-green (where the dew falls heavily)
29 Dilston	Deuelestune 1172	Settlement on Devils Water
48 Dingbell Hill	Vingvell Hill 1386	ON *thing-vellir* = fields of assembly
40 Dinley	Dunley 1279	Hill-clearing
2 Dinnington	Donigton 1242	Either named after Dun, or settlement of the people on the hill
29 Dipton	Depedene 1228	Deep valley
9 Dissington	Digentum 1160, Discintune 1190, Dicheston 1208	OE *dic* = ditch, moat, embankment, near which the people lived
15 Ditchburn	Dicheburn 1236	OE *dic-burna* = burn by a ditch or wall
36 Doddington	Dodinton 1207	Either named after Dudda, or a rounded hill
28 Dotland	Dotoland 1160	OD *Dota, Dote*: Dot's land
46 Downham	Dunum 1186	OE *dunum* = at the hills
15 Doxford	Dochesefforde 1230	OE Doc's ford
4 Druridge	Dririg 1242	OE dry ridge, sand dunes
30 Dryburn	Drieburn 1182	Stream that soon dries up (M)
37 Duddo	Dudehou 1207	OE Dudda's *hoh* = projecting ridge, or by a hill
10 Duddo	Dudden 1242	OE Dudda's *denu* = valley
28 Dukesfield	Dekesfeud 1255	OE Ducc's field (M)
6 Dunsheugh	Dunchehou 1310	OE Dunn's *hoh* = crag
30 Duns Moor	Donnismore 1479	Dunn's Moor
6 Dunstan	Dunstan 1242	OE *dun-stan* = stone hill
7 Dunstanburgh	Dunstanburgh 1321	Stone-hill fort
29 Dunstanwood	Dunstanwode 1268	Wood by the rocky hill
10 Eachwick	Achewic 1160	OE *aecen-wic* = oats farm, or *Aeca-wic* = Aeca's farm
35 Earle	Yherdhill 1242	OE *geard* = yard, hill with an enclosure, or *gerd* = rod: where rods came from
53 Earlside (now Foulshields)	Yerlesset 1200	earl's seat, or high place
2 Earsdon	Erdesdon 1233	OE Eanred's or Eored's *dun* = hill
4 Earsdon	Erdisduna 1198	OE Eanred's or Eored's *dun* = hill
16 Easington	Yesington 1242	If the ME Yese is the name of the stream, it refers to the settlers there
16 Edderacres	Edredakers 1314	Aetherdred's fields (M)
11 Edington	Ydinton 1196	OE the settlement of Ida's people

PLACE NAMES AND FIELD NAMES OF NORTHUMBERLAND

13 Edlingham (injam)	Eadwulfincham 1040	OE *Eadwulfingaham* = settlement named after Eadwulf
14 Eglingham (injam)	Ecgwulfincham 1040	OE Egwulf's named settlement
10 Eland, Little	Parva Elaund 1242	OE *ealand* = river island
16 Elford	Eleford 1256	Either Ella's or eel ford
33 Elilaw	Ylylawe 1290	OE *illa-hlaw* = Illa's hill
15 Ellingham (injam)	Ellingeham 1130	OE Ella's people's settlement
4 Ellington	Elingtuna 1166	OE Ella's named settlement or eel stream/land
54 Ellishaw (sher)	Illishawe 1254	OE *illa-scaga* = Illa's shaw, copse
39 Elrington	Elrinton 1229	OE *elren-tun* = place of the elder trees
32 Elsdon	Eledene 1226 Hellesden 1236	OE Elli's *denu* = valley
1 Elswick (elzik)	Elstwyc 1189 Alsiwic 1204	Aelfsige's *wic* = farm
19 Eltringham (injam)	Heldringeham 1200	OE *Aelfhere-ingaham* = settlement named after Aelfthere
16 Elwick (elik)	Ellewich 1154	OE Ella's farm
7 Embleton	Emlesdune 1200	OE *emel-dun*. It could be caterpillar hill, but Aemele's hill is likely
28 Embley	Elmeley 1359	OE elm-clearing (M)
30 Erringburn	Eriane 1479	British, related to Welsh *arian* = silver.
29 Errington	Erienton 1202	The farm on the Eriane stream.
38 Eshells	Eskeinggesceles 1160	OW Scandinavian *eski* = ash tree. A hill pasture with ash trees
4 Eshott (eshert)	Esseta 1186	OE *aesc-sceat* = ash-grove
25 Eslington	Eslington 1163	Esla's named settlement
28 Espershields	Estberdesheles 1230	East Burntshield (burnt hut or pasture)
12 Espley	Espeley 1242	Aspen wood
26 Etal (ee)	Ethale 1232	Either OE Eata's *halh* (haugh) or *ete-halh* = a grazing pasture
36 Ewart (uwart)	Ewurthe 1218	OE *ea-worth* = enclosure on a river
3 Ewarts Hill	Heworth 1202	OE *ea-worth* = enclosure on a river
22 Ewesley Burn (oozly)	Oseley 1286	OE *osle* = blackbird, *leah* = wood
44 Fairhaugh	Fairhaluh 1245	OE fair haugh
21 Fairnley	Farniley 1271	OE *fearing-leah* = ferny clearing, grove
7 Fallodon	Falewedune 1233	OE fallow hill
29 Fallowfield	Faloufeld 1296	Either OE *fealu* = fallow, yellowish, or OE *fealg* = newly-cultivated field
22 Fallowlees	Falalee 1388	As above
51 Falstone	Faleston 1255	OE *fealu* = yellowish or OE *fag* = multicoloured. OE *stan* = stone

PLACE NAMES AND FIELD NAMES OF NORTHUMBERLAND

59 Farglow	Ferglew 1279	?
8 Farne Islands	Farne 730	See Lindisfarne: settled by people from afar
33 Farnham	Thirnum 1242	OE *thirnum-ham* = homestead at the thorns
1 Fawdon	Faudon 1242	OE *fag-dun* = multicoloured hill
24 Fawdon	Faudon 1207	As above
21 Fawns	Faunes 1256	OE *fag-naess* = a multi-coloured ridge
59 Featherstone	Fetherstanehalg 1204	OE *fetherstan* = a cromlech with three upright stones and a capstone
37 Felkington	Felkindon 1208	OE Hill named after Feoluca
13 Felton	Feltona 1166	OE *feld-tun* = farm in large cleared fields
30 Felton Hill	Fyleton 1245	OE Fygla's settlement
11 Fencewood	Fencewood 1253	Enclosed wood (M)
27 Fenham	Fennum 1085	OE *fenn-homm* = meadow by a fen
1 Fenham	Fenhu 1256	As above
12 Fentrother	Finrode 1189	OE *fin* = heap of wood. OE *roth* = clearing
36 Fenton	Fenton 1242	OE *fen-tun* = fen farm
26 Fenwick (fenick)	Fenwic 1208	As above
2 Fenwick (fenick)	Fenwic 1242	As above
14 Filbert Haugh	Hilburhalgh 1280	OE Hildeburgh's haugh (M)
1 Flatworth	Flatforda 1271	Flatford (Dortwick sand shallows)
7 Fleetham	Fletham 1180	OE *fleot-ham* = estuary or fleet farm
36 Flodden	Floddoun 1517	OE *flode-dun* = a stream from a hill. This name does not appear until the battle
23 Flotterton	Flotweyton 1160	OE *flot-weg-tun* = farm by the raft bridge
(11) Font River	Funt 1200	Probably pre-Roman, a spring or stream
36 Ford	Forda 1225	OE ford
19 Fotherley	Falderle 1242	OE *faldere-leah* = sheep folder's meadow
39 Fourstones	Fourstanys 1236	Stones forming a 'four poster' prehistoric burial site, route markers, cromlech?
15 Fowberry	Folebir 1242	OE *folan-byrig* = defended enclosure where the foals were kept
33 Foxton Burn	Foxden 1325	Fox valley
13 Framlington, Long	Fremelintun 1166	OE Settlement named after Framela
53 Gamelspath	Kenylpethfeld 1380	OS Gemel's path
12 Garretlee	Gerardsley 1296	Gerard's clearing
42 Garret Shiels	Gerardschelels 1290	Gerard's shieling

37 Gatherick	Gateriswic 1281	? Dogwood tree farm (M)
5 Gilden Burn	Gidenes dene 1200	Gildwine's valley (M)
13 Glantlees	Glendelya 1201	OE *glente* = a lookout hill
24 Glanton	Glentendon 1186	As above
(36) Glen River	Gleni 730	Pre-Roman *glano* = clean, beautiful, holy
(36) Glendale	Glendal 1179	Glen valley
39 Glendue	Glendew 1239	Welsh *glyn* = valley, *du* = dark
5 Gloster hill	Gloucestre 1178	Could it be the same as the southern town, and mean a bright, splendid place?
40 Gofton	Goffedene 1279	OE ? Gof's valley (M)
43 Golden Pot	Goldingpot 1230	M thinks that this is the pot of Golda's people. The pots are stone blocks hollowed out on top for posts, way-markers
12 Gorfen Letch	Gorsfen 1270	OE *gorst-fen* = gorse on the marsh
1 Gosforth	Goseford 1166	OE *gos(a)-ford* = goose ford
27 Goswic (gozic)	Gossewic 1202	Goose farm
41 Greenhaugh	Grenehalgh 1325	Green haugh
59 Greenhead	Le Grenehued 1289	OE *grene-hafod* = green hill end
18 Greymare Hill	Graymere 1307 (M)	*Maere* is a boundary mark
37 Grindon	Grandon 1208	OE *grenan-dun* = green hill
29 Grottington	Grattendun 1160 (M)	Hill of Grota's people or sandy hill (*groten*)
11 Gubeon	Gobyon 1200	The Gobyon or Gubiun family left its mark on many manors
30 Gunnerton	Gunwarton 1169	Gunware's settlement (Norse)
5 Guyzance	Gysnes 1240	Guines, near Calais: Norman family name
5 Hadston	Hadeston 1189	OE *Haeddi's-tun* = farm
27 Haggerston	Agardeston 1196	?French personal name from 'haggard' - wild, odd
32 Haining	Hayning 1304	OE *haegen* = enclosure or grove
30 Hallington	Halidene 1189	OE *halig-denu* = holy valley, or enclosure?
29 Halton	Haulton 1161	OE *haw-hyll* = lookout hill farm?
49 Haltwhistle	Hautwisel 1240	OE *heafod-twisla* = place where streams join by the hill. ME *twisel*
29 Harn Burn	Hamburne 1225	OE *ham-burna* = farm by the stream
4 Hanging Leaves	Hengandelley 1262	OE *hangar* = slope. Sloping fields
15 Hangwell Law	Le Hengandewelle 1266	Spring spouting from a rock overhang

33 Harbottle	Hirbotle 1220	OE *hyra-botl* = hired persons' building (army)
35 Harehope (harop)	Harop 1185	OE *hara-hop* = hares' valley
35 Harelaw	Heyreslaw 1296	OE *hara-law* = hares' hill
19 Harlow Hill	Hyrlawe 1242	OE *higera, higre* = magpie hill, or *her(e)-hlaw* = mound of the people
21 Hartburn	Herteburne 1198	OE *heorot –burne* = stag stream
21 Harnham	Harnaham 1242	OE *haeren-ham* = stony/rocky homestead
9 Harsondale	Harestanesden 1255	OE *har-stan-denu* = grey stone valley
3 Hartford (harfod)	Hertford 1198	OE stag ford
35 Harthope Burn	Herthop 1305	OE stag valley stream
2 Hartley	Hertelawa 1166	OE *heorot-hlaw* = stag hill
58 Hartleyburn	Hertlingburne 1195	The stream of the Hartley people (M)
21 Hartington	Hertweiton 1170	OE *heorot-weg-tun* = stagpath farm
34 Hartside	Hertesheved 1255	OE stag's headland
22 Harwood House	Harewuda 1155	OE hares' wood or *har-wudw* = grey wood
28 Harwood Shiel	Harewode 1214	As above
40 Haughton	Haluton 1177	OE *halh-tun* = haugh settlement
5 Hauxley	Hauekeslaw 1204	OE *Hafoc's* (or hawk's) *hlaw* (mound)
31 Hawick	Hawic 1242	OE *hea-wic* = a high farm
39 Hawden	Hauden 1330	OE *hag-denu* = haw-valley
6 Hawkhill	Hauechil 1178	OE *hafoc-hyll* = hawk hill
51 Hawkhope (hawkp)	Haucop 1325	OE *hafoc-cop* = hawk hill
29 Hawkwell Hall	Hauekeswell 1242	OE *hafoc-wella* = hawk's spring or stream
39 Haydon Bridge	Hayden 1236	OE hay valley
26 Hazelrigg (hezlrig)	Heselrig 1242	OE hazel ridge
13 Hazon	Heisende 1170	OE *heges-ende* = end of the hedge or OE *haeg-sand* = enclosed sandy land
18 Healey	Heley 1268	OE *hea-leah* = high clearing
11 Healey	Heley 1235	As above
23 Healey	Heley 1100	As above
36 Heatherslaw	Hedereslawa 1176	OE *heahdeor-hlaw* = stag or deer hill
32 Heatherwick	Hatherwick 1250	OE heather dwelling
1 Heaton	Heton 1256	OE *hea-tun* = high settlement
37 Heaton	Hetyon 1183	As above
11 Hebron (hee)	Heburn 1242	OE *hea-byrgen*=high burial mound
14 Heckley	Hecclive 1242	OE *hea-clif* = high or *heath clif* = heather cliff

PLACE NAMES AND FIELD NAMES OF NORTHUMBERLAND

9 Heddon-on- the Wall	Heddun 1175	Heather hill
20 Black Heddon	Nigra Heddon 1242	Black heather hill
9 Heddon E and W	Hidewine 1178	OE *Hidda's-winn* = Hidda's pasture
24 Hedgeley	Hiddeslee 1150	As above
18 Hedley on the Hill	Hedley 1242	OE *heath-leah* = a clearing overgrown with heather
14 Heiferlaw	Heforside 1283	? heifer hill
12 Helm	Helm 1255	Possibly OE *helm* or ON *hialmr* = helmet, also used for a roofed cattle shed
49 Henshaw	Hedeneschalch (twelfth century)	Hethin's haugh
25 Hepburn (ebb)	(montem) Hybberndune 1050	OE *heah-byrgen* = high burial mounds
44 Hepden Burn	Heppeden 1233	OE *heop(e)-denu* = Rosehip valley
33 Hepple	Hephal 1205	OE *heop-halh* = rosehip haugh
3 Hepscott	Hebscot 1242	Hebbi's cottages or animal shelters
12 Herons Close	Heyrun 1255	OF *hairon* = heron (M)
40 Hetherington	Hetherrinton 1291	OE Haethere's people's settlement (M)
45 Hethpool	Hetpol 1242	OE Pool under or beside Hetha
26 Hetton	Hetton 1163	OE *haeth-tun* = heath dwelling
12 Heugh (Eshott-)	Hough 1276	OE *hoh* = ridge or spur
29 Hexham	Hagustaldes ea 681 Hagustaldes ham 685	*Hagustalt* = owner of an enclosure, a younger son who had to take land for himself. Water (*ea*) or island (*ei*) is replaced by homestead (*ham*)
10 Highham Dykes	Heyham 13th Century	OE *heah-ham* = high settlement
21 Highlaws	Heylaw 1250	OE high hill
11 Highlaws	Heghelawe 1293	As above
14 Hinding Burn	Henneden-burne 1275	OE *henna-dene* = hen's valley
3 Hirst	Hirst 1242	OE wood, wooded hill
13 Hitchcroft	Hitchecroft 1445	OE Hicca's croft? (M)
14 Hobberlaw, earlier Birtwell	Bertewelle 1296	OE *beorhte-wielle* = bright, clear spring
24 Holburn	Hoburn 1242	OE *hol(h)* = a deep stream, stream by a hill-spur
17 Holy Island	Halieland c 1150	Its earliest name was Lindisfarnensis 730. See Lindisfarne
33 Holystone	Halistane 1242	OE *halig-stan* = holy stone. The site of a Benedictine abbey
2 Holywell E and W	Halewell 1218	OE holy spring
16 Hoppen	Hopurn 1242	OE at the enclosures

37 Horncliffe	Hornecliffe 1208	Cliff in a *horna* = a tongue of land
19 Horsley	Horseley 1242	OE *horsa-leah* = horse pasture
12 Horseley (-Long)	Horsleg 1196	As above
2 Horton	Horton 1242	OE *horth* (or *horu*) = farm on muddy land
26 Horton	Horton Turbevill 1242	As above
10 Horton	Horton 1242	As above
9 Houghton	Houcton 1242	OE *hoh-tun* = settlement on or by the hill settlement on or by a spur
6 Houghton, Little- and Long-	Hougton Magna, Parva 1242	As aboveOE *hoh-tun* = settlement on or by the hill settlement on or by a spur
29 Houtley	Holtolaye 1243	OE *holh-leah* = hollow clearing or wood
29 Howden Dene	Holden 1290	OE *holh-denu* = deep valley
6 Howick	Hewic 1100, Hawic 1290	OE *hea-wic* = high or chief farm
46 Howtel	Holthal 1202	OE *holt-healh* = wooded haugh
32 Hudspeth	Hodespeth 1252	Hod's path (M)
14 Hulne	Holne 1283, Holen 1296	OE *holegn* = holly
35 Humbleton Hill	Hameldun 1169	OE *hamel-dun* = bare hill
30 Humshaugh	Hounshale 1279	OE Hun's haugh
★56 Huntland	Hunteland 1177	Hunter's land (M)
25 Ilderton	Ildretona 1125	ME *hilder-tun* = elder tree farm
20 Ingoe	Hinghou 1229	OE A spur called after Ing
24 Ingram	Angerham 1242	OE *angr-ham* = grassland settlement
(59) Irthing River	Irthin 1169	Pre-Roman
11 Isehaugh	Ineshaulgh 1370	Either Ina's or Isa's haugh
(17) Islandshire	Ealandscire 1170	Shire centred on Holy Island
1 Jesmond	Gesemue 1205	Mouth of the Ouse Burn (Yese = gushing)
★Karswellas	Cressewelle Leghes 1360	Fields by the cress spring (M)
20 Kearsley	Kerneslaw 1125	Cynehere's (/Cenhere's) hill or mound
48 Keenlyside	Kenleya 1230	Hill by Cena's clearing (M)
29 Keepwick	Kepwike 1279	OE Kepe's farm (M)
1 Kenton	Kinton 1242	OE *cyne-tun* = royal manor
34 Kidland	Kideland 1277	OE Cydda's land
62 Kielder	Keldre 1309	Pre-Roman, like Welsh *caled-dwfr* = violent water
46 Kilham	Killum 1177	OE *cylnum* = at the kilns
2 Killingworth	Killingwrth 1242	OE Cylla's named settlement
36 Kimmerston	Kynemerston 1244	OE Cynemaer's farm

49 Kingswood	Kingeswood 1135	OE King's wood
21 Kirkharle	Herle 1170	OE *Herela-leah* = Herela's clearing
48 Kirkhaugh	Kyrkhalwe 1254	Haugh with a church
20 Kirkheaton	Kyrkeheton 1296	High settlement with a church
10 Kirkley	Crikelawa 1176	British *cruc*, OE *hyll*, *hlaw* = hill (all three) or Crick's mound
36 Kirknewton	Niweton in Glendala 1336	OE *neowa-tun* = new homestead with church
31 Kirkwhelpington	Welpinton 1182	OE Hwelp's named settlement, with a church
68 Knaresdale (narz-)	Knaresdal 1254	ME *knar* = a rugged rock. Valley
26 Kyloe	Culeia 1195	OE *cy-leah* = cow pasture
58 Lambley	Lambeleya 1201	OE lambs' pasture
39 Langhope	Langhop 1229	OE long valley
39 Langley	Langeleya 1212	OE long wood or clearing
36 Lanton	Langeton 1242	OE long village or settlement
41 Leam	Leum 1176	OE *leum*, *leam* = at the groves/clearings
24 Learchild	Levericheheld 1242	OE Leofric's *helde* (slope)
46 Learmouth	Leuremue 1176	Mouth of the Lever (*laefer* = rush or iris)
22 Leighton, Green (lie)	Lytedon 1242	OE *leoht-dun* = light hill
14 Lemmington	Lemetun 1157	OE *hleomoc-tun* = brook lime farm (speedwell, Veronica Beccabunga)
6 Lesbury	Lechesbiri 1190	OE *laece-burg* = Physician's (leech's) manor
25 Lilburn	Lilleburn 1170	OE Lilla's burn, or the chattering stream
28 Lilswood	Lilleswrth –wude, 1233	OE Lilla's wood
40 Linacres	Linacres 1279	OE *lin-aeceras* = flax fields
17 Lindisfarne	Insula Lindisfarenensis ecclesia, Lindisfaronensis 730. Lindisfarena ea 890	Lindis is the name for north Lincolnshire. *Lindisfaran* = travellers, *ea* = island. Bede's *gens lindisfarorum* = a colony
33 Linsheels	Lynesheles 1292	The shielings by the torrent
4 Linton	Linttuna 1137	Settlement on the River Lyne
39 Lipwood	Lipwude 1176	Lippa's wood or OE *hliep*, *help-wudu* = steeply-sloping wood
13 Longframlimgton	Fremelintun 1166	Framela's named settlement
3 Longhirst (langerst)	Langherst 1200	OE Long wood or wooded hill
12 Longhorsley	Horsleg 1197	OE *horsa-leah* = horse pasture
21 Longwitton	Wittun 1236	OE *widu-tun* = wood farm
23 Lorbottle	Luuerbotl 1178	OE Leofwaru's *botl* (building)

26 Lowick	Lowich 1181	OE farm on the River Low. (Dialect *low* is a shallow, tidal pool)
27 Lowlynn	Loulinne 1208	OE *hylnn* = a pool on the river
16 Lucker	Lucre 1169	OS *lo-kiarr* = marsh where sandpipers go, or marsh pool (*luh-ker*)
26 Lyham	Leum 1242	OE *leah-hamm* = grove meadow
12 Lynch Wood	Linchwiteburn 1200	OE *hlinc* = ridge
(4) Lyne River (line)	Lina 1050	Pre-Roman: to pour, flow, glide
4 Lynemouth	Lynemuwe 1242	Mouth of the River Lyne
2 Lysdon	Lidisdene 13th Century	Lida's valley (M)
22 Maggleburn	Macgild 1261	Pre-Roman river name (M)
(28) March burn	Marchenburne 1275	?Merce's people's burn (M); on the boundary?
1 Marden	Merden 1294	OE *mere-denn* = mares' pasture or *mearc-denu* = boundary valley (M)
2 Mason	Merdisfen 1242	OE Maerheard's fen
20 Matfen	Matefen 1159	OE Mata's fen?
11 Meldon	Meldon 1242	OE *mael-dun* = hill with monument or cross
46 Melkington	Millonden 1425 (M)	?OE *mylen-denn* = pasture with a mill
49 Melkridge	Melkrige 1279	OE *meoluc-hryg* = milk ridge (good pasture)
19 Mickley	Michelleie 1190	OE *micela-leah* = large clearing
50 Middleburn	Midelburn 1286	Middle burn
16 Middleton	Middleton 1242	Middle settlement
21 Middleton	Middelton Morel 1242	As above, held by the Morel family
35 Middleton	Tres Middleton 1201	Middle settlement
10 Milbourne	Meleburna 1158	OE *mylen-burna* = mill stream
2 Milkhope	Mylkhopleche 1260	OE *meoluc-hop-leche* = milk (rich pasture) valley with watercourse
1 Milton	Mulliton 1204	Mill farm
46 Mindrum	Minethrum 1040	Welsh *mynydd* = mountain, OE *drum*, *trum* = a ridge
11 Mitford	Midford 1195	OE *myth-ford* = ford at the stream junctions
11 Moledon (Moezdn)	Molliston 1242	Moll's or Mull's farm
32 Monkridge	Munkerich 1250	Monk's ridge
1 Monkseaton	Seton Monachorum 1380	OE *sae-tun* = sea farm belonging to the monks of Tynemouth

PLACE NAMES AND FIELD NAMES OF NORTHUMBERLAND

8 Monkshouse	Broclesmouth 1257, le Monkeshouse ex parte boreali rivuli, Broxmouth 1495	From being the estuary of Brocc or Broccel, it became a storehouse for the monks of Farne
3 Moor, Old and New	Pendemor 1296	Posssibly Penda's *mor* = swamp
40 Morralee	Moriley 1279	OE *moriga(n)-leag* = swampy clearing (M)
2 Moralhirst	Mirihildhyrst 1309	OE *myr(i)ge-hylde-hyrst* = pleasant woodslope
3 Morpeth	Morthpath and Morpeth 1200	Path across the moor, but also murder path
5 Morwick (morrik)	Morewick 1161	OE *mor-wic* = fen farm
6 Mosscroft	Musecroft 1269	Mousecroft (M)
18 Mosswood	Moseforth 1378	Ford by the moss (M)
16 Mousen	Mulefen 1166	OE Mul's fen
19 Nafferton	Natferton 1187	ON *Nattfari-tun* = Nattfari's settlement
16 Nanny River	Nauny 1245	A pre-Roman name (M)
36 Nesbit	Nesebit 1242	OE *neosu-byht* = a nose-like bend
19 Nesbitt	Nesebite 1242	As above
33 Netherton	Nedertun 1050	OE *neadrdre-tun* = adder farm (W)
12 Netherwitton	Wittun 1236	OE *widu-tun* = (lower) settlement by a wood
3 Newbiggin-by-the-Sea	Niwebiginga 1187	OE new building or house
9 Newbiggin	Neubiging 1242	As above
28 Newbiggin	Neubyggyng 1378	As above
46 Newbiggin	Neubiging 1208	As above
9 Newburn	Neuburna 1121	OE new stream or borough (*burh*)
39 Newbrough	Niewburc 1203	OE new fort
1 Newcastle-on-Tyne	Novum Castellum 1130	New castle
15 Newham	Neuham 1242	New settlement
7 Newham	Neuham 1242	New settlement
16 Newlands	Neulond 1345	Either newly-acquired or newly-cleared
11 Newminster	Novi Monasterii 1137	New monastery
15 Newstead	Newstede 1377	New farmstead
2 Newsham	Neuhusum 1207	OE *(aet)neowan husum* = at the new houses
36 Newton 13 Newton	Including Kirknewton, West-, on the Moor,-by the Sea	New settlement
7 Newton		
25 Newtown	Nova villa super Warneth 1130	New settlement on the Waren Burn (= alder stream)

23 Newtown	Newtown 1248	New settlement
37 Norham on Tweed (Norram)	Northam 1040	OE northern estate. Earlier it was Ubbanford = Ubba's ford (c.1030)
Northumberland	Northhymbre 867, Northymbraland 895	People who lived north of the Humber. Used in its modern sense after c.1000
22 Nunnykirk	Graungia vocata Noonekirke 1536	Nunna's church? (W)
40 Nunwick (nunnik)	Nunnewic 1165	Nun's dwelling
12 Oakhaugh	Akehalgh 1201	OE ac-halh = oak haugh
29 Oakwood	Acuudam 1160	OE oak wood (M)
10 Ogle (owe)	Hoggel 1169	OE Ocga(n)-hyll = Ocga's hill
37 Ord	Horde 1196	OE ord = point or sword of land
42 Otterburn	Oterburn 1217	OE otter stream
31 Ottercops	Altercopes 1265	OE cop = a hill
1 Ouse Burn	Jhesam, Yese 1293	OE geosan = to gush, geose = gushing river
20 Ouston	Hukeleston 1201	OE Ulfkell's settlement
48 Ouston	Ulvestona 1297	OE Ulf's or Wulf's settlement
16 Outchester	Ulecestr 1206	OE ule-ceastre = owl fort (deserted?)
13 Overgrass	Ovegares 1255	OE ofre = margin, overlooking. ME gares = triangular-shaped field
19 Ovingham (injam)	Ovingeham 1238	OE Ofa's named settlement
19 Ovington	Oventhuna 1271	As above
29 Owmers	Ulmeres 1296	OE ule-mersc = owl marsh (M)
1 Pandon	Pampeden 1177	OE Pampi's valley (M)
48 Parmentley	Parmontle 1135	Pearmain clearing (M)
46 Paston	Palestun 1176	OE Palloc's farm
12 Pauperhaugh (pop-)	Papwirthhalgh 1120	OE Papworth's haugh
3 Pegswood	Peggiswrth 1242	OE Peggs-worth = enclosure
44 Philip	Fulhope 1331	OE foul valley (M)
11 Pigdon	Pikedenn 1205	OE Pica-denn = Pica's pasture, or pointed hill (W)
33 Plainfield	Flaynefeld 1272	Possibly Fleinne's field (M)
48 Plenmeller	Plenmenewre 1256	Welsh blaen moelfre = top of the bare hill
2 Plessey	Pleisiz, Pleistum 1203	OF plaisseis, plaissiet = park enclosed by plashed hedge (interwoven, bent branches)
14 Plunderburn	Plundenburne 1220	?plum stream (M)
10 Pont River	Ponte 1269	Welsh pant = a valley
10 Ponteland (ee)	Eland 1242, Punteland (1203)	OE egland, ealand = an island (in marsh), or newly-cultivated land by the River Pont (W)

PLACE NAMES AND FIELD NAMES OF NORTHUMBERLAND

29 Portgate	Portyate 1269	Gate through the Roman Wall
24 Powburn	Powburn 1868	A pit or pool in a stream (*poll*)
*51 Powtreuet	Poltrernerth 1325	Pre-Roman (M)
24 Prendwick	Prendewick, Prendwyc 1242	OE Prenda's farm
46 Pressen	Prestfen 1177	OE the priest fen
15 Preston	Preston 1242	Priest's estate
1 Preston	Prestona 1198	As above
10 Prestwick (tik)	Prestwic 1242	Priest's farm
9 Prudhoe (prudda)	Prudho 1173	OE Pruda's *hoh* = projecting ridge
33 Puncherton	Pun(t)chardon 1250	Punchardon was the Norman French owner
49 Ramshaw	Ramschawes 1312	OE *hraefnes-wudu* = raven's wood (M)
53 Ramshope	Rammeshope 1230	OE *hramsa-hop* = wild garlic valley
15 Ratchwood	Wretheswode 1279	OE *wrecca-wudu* = outlaw's wood
40 Ravensheugh	Ravenshugh 1354	OE raven's *hoh* (ridge)
31 Ray	Raye 1300	M says that this is either *ray* = darnel or *wray* = landmark. It could also be *aet thaere ea* = at the river
32 Raylees	Raleys 1377	OE roe-deer clearings ? (M)
24 Reaveley	Reueley 1242	Reeve's or rough clearing (*hreof-leah*)
49 Redburn	Redburn 1255	OE *reod-burna* = reed-burn (M)
41 Rede River	Rede 1200	OE *reade* = the red one
41 Redesdale (ridz)	Redesdale 1075	The Rede valley
41 Redeswood	Rode- Rede-wode, 1255	Rede wood
59 Redpeth	Redepeth 1255	Red or reed path
6 Rennington	Reiningtun 1104	OE Regna's named settlement
19 Riding	Ryding 1262	OE *ryding* = a clearing
19 Riding Mill	Ryding 1262	Mill in a clearing
14 Ridlees	Reddeleys 1320	Possibly cleared land (M)
23 Rimside Moor	Rimescid 1268	On the rim of (the Millstone Burn)
11 Riplington	Riplingdon 1242,-tone 1251	Settlement of the Riplingas people.
22 Ritton	Rittona 1145	OE *rith-tun* = small stream settlement
42 Rochester	Bremenio c150 Roff 1208	Bremenium is fort by the roaring stream. Hrofi's fort or rook's fort (OE *hroc*).
7 Rock	Roch 1164	OE *rocc* (OF *roche* and *roke*)
25 Roddam	Rodun 1201	OE *rodum* = at the clearings
*1 Rodestane	Rodestane 1320	OE cross-stone (M)
15 Roseborough	Osberwick 1252	Osburh's farm (M)

67

25 Roseden	Russeden 1242	OE *rysc-den* = rush valley
25 Ross Castle	Rosse 1208	Welsh *rhos* = moor, Irish *ros* = hillock, promontory. It is both. Prehistoric fort.
23 Rothbury	Routhebiria 1125	OE Hrotha's *burg* = Hrotha's fortification or the red fort (ON *rauthr*) (W)
21 Rothley	Rodolie 1086	OE *roth(u)-leah* = clearing, open ground (Ek) OE Hrotha's clearing (M)
40 Rouchester	Rowchestre 1348	OE fortification in rough ground
13 Roughley Wood	Ruely 1296	Rough clearing
44 Rowhope	Ruhope 1233	OE rough valley
9 Rudchester	Rodescastre 1251	OE Rudda's fortification (old Roman fort)
14 Rugley	Ruggele 1210	OE *hrucge-leah* = woodcock meadow
20 Ryal	Ryhill 1242	OE rye hill
24 Ryle	Parva Rihull 1212	OE rye hill
29 St John Lee	Capella Beati Johannis de Lega 1310	Church of St John in the clearing
10 Saltwick	Saltwyc 1268	Place where salt was made (M)? or farm among the sallows
29 Sandhoe	Sandho 1225	Sandy spur, ridge
1 Sandyford	Sandeforthflat 1384	Sand ford or sandy-island-ford (OE *eg*)
33 Scrainwood	Scravenwod 1242	OE *screawena-wudu* = wood of the shrewmice or villains; *scraefen-wudu* = wood at the hollow place (W)
27 Scremerston	Scrimestan 1130 Schermeretun 1196	Skirmer's boundary stone or estate (W) *Schermer* = 'scrimer' = fencer (French)
6 Seaton	Seyton 1280	OE *sae-tun* = homestead by the sea
3 Seaton North	Seton 1242	As above
2 Seaton Delaval	Seton de la Val 1270	As above, held by the de la Val family
2 Seghill	Sihala, Syghal 1271	OE *Sige-halh* = haugh on the Sige stream?
★ 28 Sessinghope	Sessynghop 1336	Cissa's valley (M), but could it be a Normanised form of Seaxinghope?
39 Settlingstones	Sadelingstan 1255	A place where people mounted their horses
40 Sewing Shields	Swyinscheles 1279	OE Sigwine's shielings
3 Shadfen	Shaldefen 1257	OE *scealda-fen* = shallow fen, in a hollow
21 Shaftoe	Shaftho 1231	OE *scealf-hoe* = shaft-shaped ridge
33 Sharperton	Scharberton 1242	OE *scearp-beorg* = steep or pointed hill farm

24 Shawdon	Schaheden 1232	OE *scaga-denu* = copse valley
3 Sheepwash	Sepewas 1178	OE *sceapwaesce* = sheepwash
11 Shelly	Shelyngley 1290	OE clearing with a shieling
28 Shield Hall	Schelis 1296	ME *schele* = temporary herd's hut
1 Shields, North	Nortscelis 1273	Temporary huts (for fishermen)
1 Shields, South	Suthshelis 1313	As above
13 Shield Dykes	Swynleys 1288, Swynleysheles 1314	Swine clearing, then shielings nearby, then a dike (wall or ditch) nearby
1 Shieldfield	Schenefeud 1255	Fields with shieling (M)
13 Shilbottle	Siplibotle 1228	OE *Scipleaingas-botl* = buildings on the estate named after Shipley
43 Shilmore	Shouelmore 1292	Shovel-shaped moor or steep hill moor (W)
11 Shilvington	Schilington, -don 1242	OE *scylf, scelf* = rock, crag where people lived (M sees it as Scylla's people)
14 Shipley	Schepley 1236	OE *sceap-leah* = sheep pasture
42 Shitlington	Sutlingtun 1240	Scyttel's named settlement
42 Shittleheugh	Schotelhough 1378	? Scot's haugh (M)
19 Shoreston Brae	Schortendene 1290	Short valley (M)
8 Shoreston	Schoteston 1177	OE Scot's settlement
37 Shoresworth	Scoreswurthin 1085	OE *score-worth* = steep slope enclosure
21 Shortflatt	Le Scortflat 1284	OE short flat or furlong
18 Shotley	Shotley 1242	OE *sceot-leah* = pasture on a steep slope
16 Shotton	Sothune 1196	Scot's settlement (M)
46 Shotton-in –Glendale	Scotadun 1050	Hill of the Scots
40 Simonburn	Simundeburn 1229	OE Sigemund's stream
22 Simonside	Simundessete 1279	OE Sigemund's (*ge*)*set* = seat or settlement
58 Slaggyford	Chaggeford 1218	ME *slag* = a ford slippery with mud or OE *ceaga* = brushwood (W)
28 Slaley	Slaveleia 1166	OE *slaef-leah* = muddy piece of cleared land
3 Sleekburn	Sliceburne 1050	Muddy stream (dialect *sleech, slitch, sleek*)
51 Smales	Smale 1279	OE *smael* = narrow
51 Snapdaugh (Snapduf)	Snabothalgh 1325	OE *snab-halh* = haugh by the little hill (M), or *snaep* = boggy
14 Snipe House	Swinleysnepe 1290	Possibly swine clearing or grazing
23 Snitter	Snitere 1176	ME *sniteren* = to snow (Dialect *snitter* = a biting blast) An exposed, cold place
13 Snook Bank	Schakelzerdesnoke 1264 Skalkelyerdesnoke 1273	A shackle-yard where the cattle were tied, on a sharp pointed projection of land

48 Snape	Suanhope 1325	Boggy valley?
32 Soppit	Sokepeth 1292	OE *soc-peth* = a marshy path
*Sowerhopeshill	Suggariple 1050	Cheviot location and meaning not known
16 Spindleston(e)	Spilestan 1165	OE *spinele-stan* = stone pillar like a spindle
53 Spithope	Spithope 1324	A spit-shaped valley
27 Spital Hall farm	The Spitle 1695	ME *spitel* = hospital (for lepers here)
6 Stamford	Stauford 1242	OE stony ford
20 Stamfordham	Stanfordham 1188	Settlement at the stony ford
2 Stannington	Stanigton 1242	OE *stanweg-tun* = paved road settlement
12 Stanton	Stantuna 1200	OE *stan-tun* = stone farm
39 Staward Peel	Staworth 1215	OE *stan-worth* = stone enclosure
28 Steel	Le Stele 1269	Precipice (dialect *steel* from OE *stigol* = stile, but later used to denote a steep ascent)
19 Stelling	Stelling 1242	Dialect *stelling* = cattle shed or enclosure
4 Stobswood	Stobbeswod 1250	OE *stobs-wudu* = a wood with tree stumps
19 Stocksfield	Stokesfeld 1242	OE *stoc-feld* = field belonging to a religious house or outlying hamlet (W)
39 Stonecroft	Stancroft 1175	OE stone croft
16 Stotfold	Stotfald, Stodfald 1244	OE *stod-fold* = stud fold
5 Sturton	Stretton 1242	OE *straet-tun* = on a paved road (Roman?)
18 Styford	Styfford 1212	OE *stig-ford* = path ford
41 Sunday Burn	Sunday-burn 1291	?
41 Sundaysight	Sunday heugh 1325	?
7 Sunderland, North	Suthlanda 1176	OE South newly-cultivated or detached land
16 Swainston	Swayneston 1351	Sveinn's farm (M)
13 Swarland	Swarland 1242	OE *swaer, swar-land* = heavy ploughland
31 Sweethope	Swethop 1215	A sweet, pleasant valley (good pasture)
30 Swinburn	Swineburn 1242	Swine stream
7 Swinhoe	Swinhou 1242	Swine ridge
52 Tarsett	Tyreset 1244	Possibly *Tir-saet* = Tir's place (M)
40 Tecket	Teket 1279	A pre-Roman name (M)
39 Tedcaster	Tadecastell 1364	Tada's fortification (M)
40 Tepper Moor	Tepermore 1479	?
59 Thirlwall	Thurlewall 1256	OE *thyrel* = perforated. A gap in the Wall

13 Thirston	Thrasfriston 1242	OE Thrasfrith's settlement. Possible nickname: ON *thrasa* = threaten ('pushy')
19 Thornborough	Thorneburg 1242	OE *thorn-burh* = a fort where thorns grew
49 Thorngrafton	Thorgraveston 1150	OE *thorngraf-tun* = thornbrake settlement
37 Thornton	Thornetona 1208	Thorn bush settlement
21 Thornton, E and W	Torinton 1203	Thorn bush settlement
12 Thornyhaugh	Thornihalugh 1309	OE a haugh overgrown with thorns
38 Threepwood	Trepwoode 1308	OE *threapian*, ME *threpen* = disputed owner
9 Throckley	Trocchelai 1161	OE *Throca-hlaw* = Throca's burial mound
30 Throckrington	Thokerinton 1223	Thoker's named settlement
11 Throphill	Trophill 1166	OE farm hill
23 Thropton	Tropton 1177	It might be the farm at the crossroads (Ek) or an estate with an outlying hamlet (W)
23 Thrunton	Trowentona 1180	?Thurwine's farm (M)
(36) Till River	Till 1040	Pre-Roman (M). Ek identifies it with the Tille in France, meaning to dissolve, flow
47 Tillmouth	Tylemuthe 1050	Mouth of the River Till
24 Titlington	Tedlintona 1123	OE Settlement named after Tytel or Tytta
12 Todburn	Todborne 1434	Fox stream (or Toda's)
48 Todhill	Todholes 1312	Fox holes
30 Todridge	Todrige 1479	Fox ridge
5 Togston	Toggesdena 1129	Tocg's valley
31 Tone	Tolland 1296	? OE *toln-land* = land on which a toll is paid
23 Tosson	Tosse, Thosse 1150	OE *tot-stan* = a lookout stone
11 Tranwell	Trennewell 1267	ON *trani* = crane's stream, spring
23 Trewhitt	Tirwit 1150	ON *tyri* = dry, resinous wood. OE *thwit* (thwaite) = clearing in the forest. It could be a wooden-built farm
10 Trewick	Trewyc 1242	OE *treo-wic* = tree farm
25 Trickley	Trikelton 1177	? sheep dung farm (M)
4 Tritlington	Turthlyngton 1170	OE Tyrhtel's named settlement
42 Troughend (trufend)	Trocquen 1242	Possibly OE *thruh* = settlement at the pipes or troughs (W)

PLACE NAMES AND FIELD NAMES OF NORTHUMBERLAND

44 Trows	Wytetrowes 1197	White troughs/depressions? (M) or trees?
7 Tughall	Tughala 1104	OE *Tugga-healh* = Tugga's haugh
(27) Tweed River	Tuidi fluminis 730, Tweoda 1040	Pre-Roman, possibly meaning powerful (Ek)
27 Tweedmouth	Tuedemue 1208	Mouth of the Tweed
15 Twizel	Tuisele 1208	OE *twisla* = the fork of a river, junction
10 Twizell	Twisle 1050	As above
(1) Tyne River	Tinea 700	From *ti* = to dissolve, flow. OE *thinan* = to dissolve? (A similar root for the river Till)
(29) Tynedale	Tindala 1158	Tyne valley
15 Tyneley	Tyndeley 1278	Possibly like Tyne above, but a smaller flow
1 Tynemouth	Tinanmuthe 792	Mouth of the River Tyne
4 Ulgham (uffm)	Wlacam 1139, Ulweham 1242	OE *ule-hwamm* = owl corner/nook
59 Ulwham	Ulgheham 1479	As above
24 Unthank	Unthanc 1207	OE *unthances* = without leave (squatters)
49 Unthank	Unthanc 1200	As above
39 Vauce	Vaus 1329	A Norman name
1 Walker	Waucre 1242	OE *wall*, ME *kerr* = marsh near the Wall
29 Wall	Wal 1166	On the Roman Wall
9 Walbottle	Walbotl 1176	OE *botl* = building on the Roman Wall
21 Wallington	Walington 1242	W(e)alh's people's settlement
1 Wallsend	Wallesende 1085	The end of the Roman Wall
59 Walltown	Waltona 1279	Settlement on the Roman Wall
30 Walwick	Wallewik 1262	OE *wic* = farm on the Roman Wall
(3) Wansbeck River	Wenspic 1137	?OE *waegn-spic* = a wagon-brushwood causeway (Ek)
29 Warden (West)	Waredun 1175	OE *weard-dun* = watch (lookout) hill
15 Waren Burn	Warnet 1212	British *verno* = alders
15 Warenford	Warneford 1256	Ford on the Waren Burn
15 Waren Mill	Molend de Warnet 14th C.	Mill on the alder stream (W)
16 Warenton	Warnetham 1209	Settlement on the Waren Burn
40 Wark	Werce 1279	OE *(ge)weorc* = fort
56 Wark	Werch 1157	As above
5 Warkworth	Werceworthe 1040	OE *Werce-worthe* = Werce's settlement. An abbess of Tynemouth in the seventh century was called Verca
18 Waskerley	Waskerley 1262	OE *waesse-leah* = marsh clearing

23 Warton	Wartun 1236	OE *weard-tun* = watch/lookout place
1 Weedslade	Wideslade 1197	OE *withig-slaed* = willow/withy valley
25 Weetwood	Wetewude 1197	Wet wood
12 Weldon	Welden 1250	OE *wielle-denu* = spring valley
11 Wharlton	Walton 1203	OE *hwae-tun* = hill farm
39 Wharmley	Quarenley 1279	OE *cweorn-leage* = mill clearing
2 Wheatridge	Whytrig 1296	OE *hwit-hrycg* = white ridge
5 Whirleyshaws	Qwirlecharr 1350	quarry ridge or wood
39 Whinnetley	Wineteleia 1207	a field, but name obscure
34 Whiteburn	Whiteburne 1233	white burn (M)
39 Whiechapel	Whitchapel 1368	white chapel (M)
2 Whitehall	Wytelawe 1250	white hill
9 Whitchester	Witcestre 1221	OE *wyt-ceaster* = white Roman fort
48 Whitfield	Witefeld 1254	White field
2 Whitley	Wyteleya 1198	OE *wyt-leah* = white meadow
24 Whittingham (injam)	Hwintincham 1040	Hwita's named settlement
20 Whittington	Witynton 1233	Hwita's named settlement
13 Whittle	Wythill 1266	OE white hill
19 Whittle	Wythill 1242	OE white hill
23 Whitton	Witton 1228	As above, or Hwita's or wood settlement
18 Whittonstall	Quictunstal 1150	OE *cwichege-tunstall* = farm with a quickset hedge
21 Whittonstone	Le Whystan 1292	The whetstone (M)
58 Whitwham	Le Whitewham 1317	OE *wyt-hwamm* = white valley or corner
9 Whorlton	Wheruel-, Wherwelton 1323	OE *hwerfel-tun* = farm by the round hill
4 Widdrington	Vuderintuna 1160	OE Widuhere's named settlement
33 Wilkwood	Wilkewde 1230	OE Willoc's wood (M)
49 Willimontswyke	Willimoteswike 1279	(French) Willimot's farm
1 Willington	Wiflintun 1107	OE Wifel's named settlement
44 Windyhaugh	Windihege 1200	ME *hege* = windy hedge
12 Wingates	Wyndegates 1208	OE *wind-gate* = pass where the wind blows
21 Witton, Long	Witun 1236	OE *widu-tun* = wood settlement
12 Witton, Nether	Wittun 1236	Lower wood settlement
41 Woodburn	Wodeburn 1265	Stream coming from the wood
5 Wooden (woodn)	Woveden 1265	OE wolves'o hill (M) or valley (Ek)
3 Woodhorn	Wudehorn 1177	OE wooded point of land

35 Wooler	Wulloure, Welloure 1187	OE *wella-ofre* = spring promontory, margin
10 Woolsington (wisintn)	Wulsinton 1203	OE Wulfsige's named settlement
25 Wooperton	Wepredane 1180	Possibly OE *weoh-beorg-denu* = temple-hill-valley (Ek)
33 Wreighill (ree)	Werihill, Werghill 1293	OE *wearg-hyll* = felon hill?
23 Wreigh Burn (rye)	Rye 1540	OE *wearg* = a felon (M)?
59 Wydon	Wyden 1255	Wide valley (M)
9 Wylam	Wilum, 1198, Wylum 1271	OE *wil* = a mechanical device, or a fish trap, and this could be a water mill on a *hamm* = water meadow
33 Yardhope	Yerdhopp 1324	Valley marked by an enclosure (M)
44 Yarnspath Law	Hernispeth 1233	OE *earnes-peth* = eagle's path
29 Yarridge	Jernerig 1232	OE *gearwe-hrycg* = yarrow-grass ridge (M)
41 Yearhaugh	Yarhalgh 1312	Fishery-haugh (M)
36 Yeavering (yevrin)	Adgefrin, Adgebrin 731	Welsh *gafre* = wild goat. Gefrin is the old name for Yeavering Bell hillfort on a bell-shaped hill
23 Yetlington	Yettlinton 1186	OE Geatela's named settlement

12 From Humbleton Hill over Monday Cleaugh to Yeavering Bell

PART II

FIELD NAMES

GENERAL REVIEW

Place names are constantly being reviewed, and changes in derivation and meaning are occasionally made, but the amount of material for field name study is enormous. There are literally thousands to consider and there are thousands more that have disappeared.

For me, the discovery of some field name maps has been accidental. When I lived in Felton I had managed to fill in many of the mapped fields with names of different periods, yet with many gaps left. News of my hobby spread, and one day I was given the 1772 survey of some farms within the Riddell estates. They were drawn in ink on paper stuck down in the pages of a book on linen. Having recorded them, I presented them to the Record Office in Gosforth on behalf of the Blundell-Brown family.

Another chance came after a talk in Corbridge, when a member of the audience offered me maps in her possession which related to the sale of The Tyneside Estate in 1873.

I have on many occasions been disappointed to find that the numbers do not have a list of names with them, but here there was a catalogue printed by the auctioneers, Messrs Driver at 5 Whitehall, London, with every field listed and its 'quantity' in acres, roods and perches given. There were 11 lots in all. The discovery came long after I had written my 1977 book on field names, and rekindled my interest, partly leading to the book that you are now reading.

Such a map covers lands administered by Greenwich Hospital, which was ordered to administer them by the Crown after the execution of the Jacobite Earl of Derwentwater. The surveyors of these estates had produced high quality maps in the early 1800s, with names of fields included. Many farms lay far outside the Tyne valley, such as Scremerston, Spindlestone and Outchester. Collectively, these maps are a wonderful record and one can read them like a book.

Another fortunate coincidence for me was that I worked at Alnwick College of Education, training teachers, for 11 years until its closure in 1977. I had received great help from the Duke of Northumberland and his estates staff in my work on prehistory through the castle museum and its catalogues, and when my mind turned to the origin and meaning of names, I had there one of the most detailed and best-preserved sources in the North. My research only scratched the surface of what is available, but it provided me with insights into how names changed over time and the importance of what they tell us about the past. The Tithe maps of the mid-nineteenth century are very detailed, showing not only the size and shape of fields but their names. The earlier maps, especially those works of art created by Robert Norton, are painted with all the strips of different kinds of land holding and their different uses; the names of fields and sizes are carefully inked in.

Another source is from farmers whom I approached and sometimes from students taught by me. For example, Mrs Sanderson (and her husband) owned Eshott farm when

I lived in Felton, and she attended evening literature classes that I taught. When they realised that I was researching field names, they invited me to their home and showed me the maps that they had. Two were gems, because, although they were modern, they showed how names can change over a 100-year period. Three in particular stand out, and I use them to show how perilous the process of studying names can be. On the 1877 map there is a field called The Tofts; on the 1976 map it had become High Seas. The reason? The owners stood one day to watch the wind rippling the surface of wheat. Sheer poetry! A field called South East Fen in 1877 was changed to March Brown in 1976. Why? Colonel Brown was a military man who marched rather than walked. This was part of his usual route. Now who would have guessed that? Quality Walk was for people of 'quality'. One more, this time from a survey by Pat Wilson of Slaley, who did an evening history course with me at the local high school and chose to research her local field names as a special study. She found a field called Seagull Field, where a one-legged seagull used to sit on a gate and watch the work going on in the field.

These examples illustrate the intimate, personal factors at work in naming the land, and should make us humble in our approach. It is possible that we are totally wrong about meanings.

There are so many sources to explore; the further back we go, the more we can appreciate which, and possibly why, names survive. Not all are very exciting: modern names can cause us to yawn when we find many directional words and not very much variety. The unusual tend to be the most challenging. To take a few examples of names with interesting sounds that I particularly like and often quote, for there is an aesthetic of field name study: Grimping Haugh, Cork Leach Lay, Hot Bog, Niddles, Armourer's Fall, The Boiling Riggs, Bodle Hole Quarter, Blakehopesburnhaugh, Crum Roods, Fiselbee Pasture, Corney Horners, Hungerfull lases, Haughslopp butts, Moralees, Seggy Hole and East Toddles. These are taken from a variety of maps and you might like to guess what they mean.

Modern agriculture has decreed that fields should be large, so that machines can more effectively plough, sow and harvest crops. Miles of hedgerows have been ripped up, the loss of which has become so alarming that funds have been made available to replant some of them. Intensive farming has destroyed the habitats of some wild creatures and we now realise what we have lost. Supermarkets too often dictate what is to be grown, and of what size and shape. The use of chemicals has polluted streams and rivers and endangered our health. Those who farm have become a very small minority of the working and dependent population; the countryside is seen by some as an outlet for leisure rather than a productive asset. We neglect farming at our peril.

These recent changes, alarming though some of them are, are not uniform, and Northumberland has fortunately been protected against the creation of universal prairie-like fields by the nature of its landscape and by the care and interest of its owners. We think of fields as areas enclosed by walls, fences and hedges, but the earliest were large open clearances in wooded land. In the Cheviot Hills they were constructed as terraces on the hillsides, are probably in some cases of prehistoric origin, retaining soil and moisture as they do in the Mediterranean regions today. The large open-fields, numbering from one to four in a collective settlement, were divided into strips and a system emerged in which people farming the land shared it by being allocated strips in different fields. In this way they could share the best and the worst of the land. As ploughs became more sophisticated, they could share ownership and the task of guiding the oxen pulling them. Each strip, roughly a furlong in length, built up in time as the soil was forced inward. Furrows were left between rigs, and these helped to drain the land; whether farmers knew it or not, a corrugated surface retained more warmth than a flat surface. The strips look like inverted Ss, as the plough drawn by yokes of heavy oxen had to swing round at the headland when moving up or down the slope or on level ground. We can see these clearly in the Cheviots in places like the Ingram valley, where they cut across earlier lynchets and terraces. The arable was abandoned for grassland (thus the name 'Ingram'), and retains these fossilised early farming patterns. Not only in the hills, but almost everywhere in Northumberland there are examples of rig and furrow of different periods, including narrow strips in much later enclosed fields, and steam-ploughed wide, regular rigs. The later rig and furrow cultivation is not based on the strip system of medieval fields, for by then the social and agricultural system had changed.

The old rig and furrow is best seen when the sun is low in the sky, or when there is light snow or frost. There are a number of 'deserted' medieval villages where the pattern

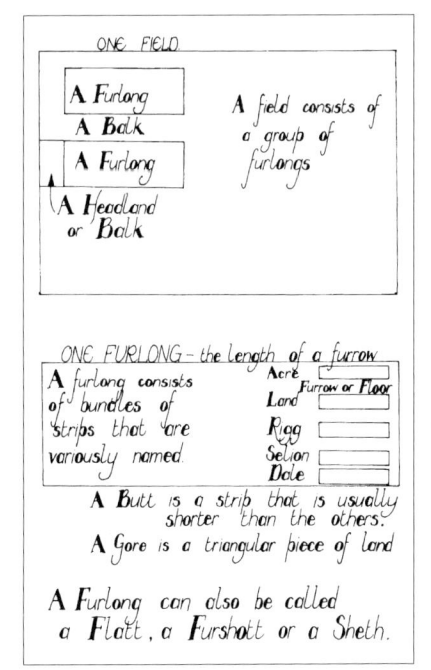

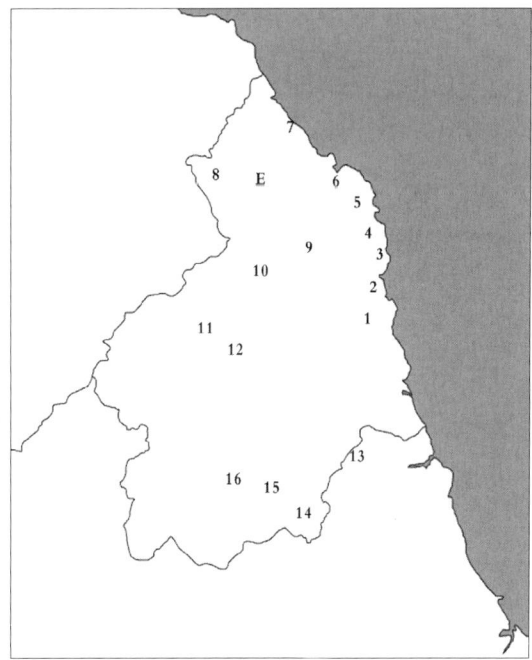

Above left: 13 Diagram of parts of the large fields

Above right: 14 Map of field name locations: E – Ewart; 1 Felton, Thirston, Eshott, Acklington; 2 Shilbottle, Buston, Warkworth; 3 Lesbury, Bilton; 4 Longhoughton, Rennington; 5 Ellingham, Newham, Tughall; 6 Lucker, Detchant, Spindlestone, Outchester, Bamburgh; 7 Scremerston; 8 Pawston; 9 Alnham; 10 Rochester; 11 Otterburn; 12 Eldson; 13 Throckley; 14 Whittonstall, Newlands; 15 Dilston; 16 Coastley

of cottages bordering the main street and rig and furrow systems all around have survived. These are particularly clear from the air, but many can be appreciated from the ground.

Fortunately we have maps in Northumberland of the late sixteenth and early seventeenth century, hand painted, that show the old system at work, in that all the rigs are there under different terms of ownership and so are common pastures. The diagram simplifies how the fields were divided in the 'open-field' system.

The clearest picture is from the Robert Norton maps. He surveyed the Earl of Northumberland's land holdings not only in considerable detail, but created beautiful paintings of them. Accompanying them, in immaculate script, are detailed records of who farmed each strip.

Remnants of ancient field systems are more likely to be visible in poorer upland areas than in the lower parts that have been ploughed over many times. 'Strips' included lynchets, and in parts of England the Romans used strips about 100m wide.

In the open-field system, the diagram shows how fields were made up of furlongs, which were further subdivided into bundles of strips separated by furrows and baulks. The baulks gave access to different strips and some developed into farm tracks, later to be retained as winding roads, contrasting with the straight stretches of roads built during the enclosure of common land and the building of turnpikes, like parts of the Hexham–Alnmouth 'Corn Road'.

The origin of strips was thought to be a result of sharing ox-teams, but this is no longer considered true. Those who cleared the land shared the enclosed parcels, a system which continued in Northumberland into the seventeenth century. 'Assart' lands (cleared from forest and waste) were parcelled out, and so were 'Demesne lands' ('Mains' on field maps) when they became available. This system came from the Anglian settlement, but enclosures generally took over from around 1550.

Rigs and lands are terms often used for the same feature. Like selions, they are 0.5-acre bands, separated by trenches and bunched in pairs to give strips of 1 acre. In theory one 'acre' could be ploughed in one day by an ox-drawn plough, but the time taken would depend on what kind of land it was.

When we look at the patterns of land holding in the fields, we see strips bundled into furlongs, which are not parallel but run in different directions, often in a reversed 'S' pattern. Odd shapes and bundles are often left between the furlongs, called gores or butts ('butt-end' is still used for the end of a cigarette). The strips were separated by baulks, some of which became lanes, and some were overgrown (e.g. Oaky Baulks). The open fields themselves were fenced to keep out animals and the furlongs were sometimes hedged too.

The pattern that we see is accentuated by the corrugated effect of ploughing away from the neighbouring strips, thus increasing the height of the central rig. It is easy to generalise about field patterns, but we must remember that each region was different; farming on the coastal plain was different from farming on the hills or scarps; soil conditions varied. The more fertile areas favoured the growth of nucleated villages and towns. Yet, no matter what system was adopted, all the fields were named.

The abandonment of the strip system had many causes, among which was the outcome of episodes such as the Black Death, when half the population of England was wiped out. There was not enough labour to continue farming at the same level (although there was also less demand for food), but those who survived were in a good bargaining position to get more land. Another factor on the border was the insecurity of life, and the abandonment of some plots. The old strip farming system was found to be inefficient, as it took more time and effort to move from one large strip to another, so there was a tendency to bundle the strips together into bigger holdings, sometimes by exchange. Enclosure of the strips was accompanied by encroachment into what was regarded as 'common' land. We begin to see a landscape resembling that of today; by 1750 most open and common fields had disappeared. By 1800 the field patterns looked very much as they do today, as the maps in this book illustrate.

These observations are not an in-depth study of agriculture (which is far more complicated), but a background to the field names.

Why was it necessary to name fields? When a close relationship exists between people and the land, in which people follow a plough pulled by animals, and a system of crop rotation, manuring, sowing and harvesting, with labour-intensive methods that involve sharing, and intimate knowledge of the soil is used, it is natural to name the fields. Some may be 'lazy', producing little, others may be far away from the village and take time to get there, some may carry stories of feats of strength, of folly, of spirits that dwell there, of what grows best, of what is boggy, and so forth. Let us go back to 1479, to a Terrier (a land-list) from Chollerford on the North Tyne for examples of names given:

Kilneflate, Horselawpule, Schothalghbankys, Overshotlaubankes, Blaklaw, le Lons, Bronslauemedoue, Bronslawflate, Canonflatte, Canonbutts, Canondike, lez hevedlandes de Brouneslawflatte, Maynflatt, le Crosse, Holmersbank, Harlawhop, le Messeway, le lonynghed, Aldchestre, le Stobithorn, Morelaw, Dueldrigge, Smythehopside, Craustrige, Westraustrige, Estraustrige, Fartirmerethorne, Faltemere, lez Merlepottes, Waynrigg, Brereryg, Hudesrodes, Korhilles, Milnrig, Fulrig, Swynburne-feld.

Many of the elements are frequently present in other similar lists: flat, shott, butts, land, rigg, field. All these indicate the way the great field was split into parts. If we can trust the sound of the words to approximate to something that we recognise, a piece of level land in the common field had a limekiln on it; a mill rig had a mill. Merlepots had clay. A shott was a furlong, which included a 'heugh', land by the river. There are flats of land named Brown and Canon; there were rigs named as being foul, Dueld (dualled?), Crow's, west, east, wagon's, barley, Maynard's. One was a plough headland. Hud had some roods, the Swynburnes had a field, there was a black hill, a Brownshill meadow and a Brownshill flat. Some of the ground was wet, with a mire and a horsehill pool. There was a road (lonnen), a road head and Messeways (path through boggy ground). Being close to the Roman Wall, there was an old fort. There was a hollow bank on a river or stream, brushwood and some wasteland on a hill.

All that information came from documents. What follows comes from a map of Ewart in 1757. The mapped area lies on the Milfield Plain, where there is much archaeological evidence of settlement from the earliest people in prehistory, as well as a strong Anglian

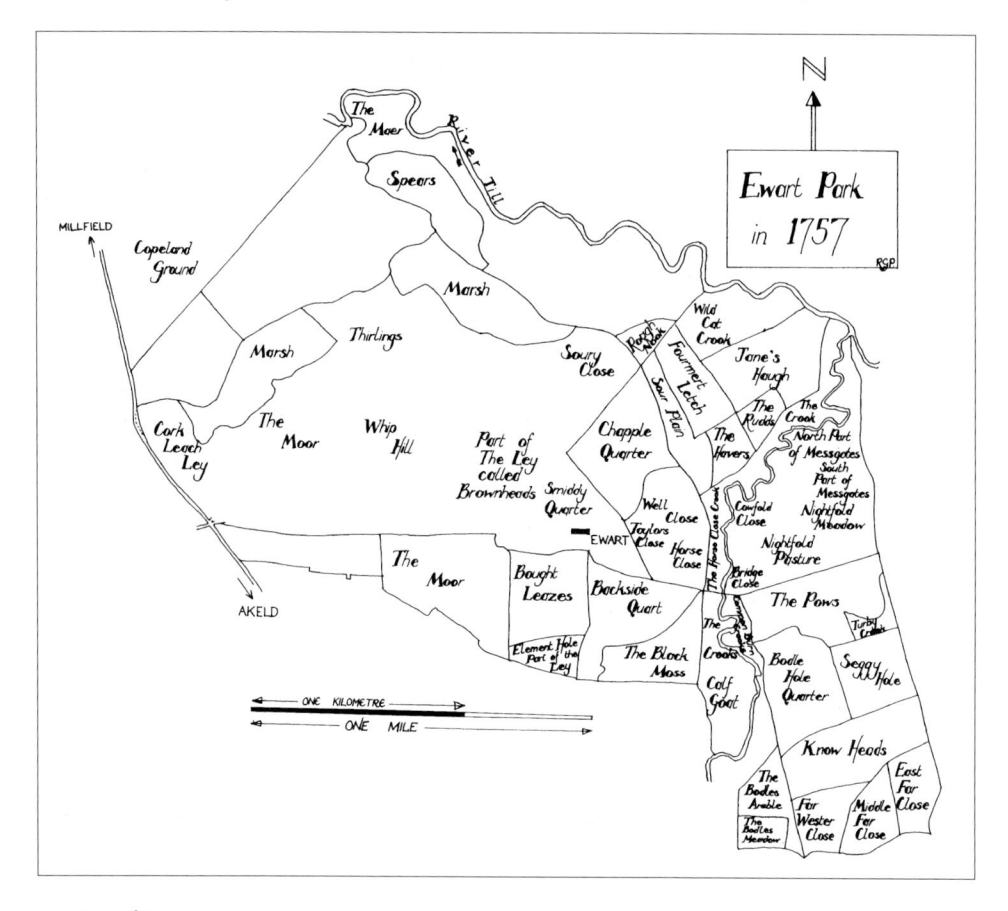

15 Map of Ewart in 1757

presence, with houses and the site of the 'palace' of Maelmin, from which the Milfield Plain got its name. It adjoins the River Till, so some names refer to bends in the river, called crooks: Bridge Crook, Common Burn Crooks, The Crooks and Crook. There were small pools and streams (The Pows). Reeds grew there (The Rudds), in parts the land was boggy with slow-moving water (Foumert Letch), dark (The Black Moss). Hazel trees or bushes may account for the name Cork Leach Ley, a meadow where a sluggish stream provided good conditions for hazel growth and material for corf-rod baskets. There was a head of a small rise (Know Heads). The Ridge and the March show where there was a boundary. There was moorland, one ploughed and the other that had never been ploughed. The size and position of fields are given: Well close (spring), South Part of the Messgates – stints of pasture in mossy ground, Little Close and North Part of the Messgates. There is a Far Wester Close, Middle Close, East Far Close, a division of land called Backside Quart, a Smiddy Quarter allocated to the blacksmith and a Chappell Quarter.

Agricultural practice is set out in Night Folds, where animals were put at night, Night Close and Cowfold Close. Other animals named are in Horse Close and Calf Goat – a 'gote' being where water goes out, usually, but not in this case, through sand or slake, to the sea. Boggy and unproductive land is indicated by Soury Plain, Rough Nook and the Moor.

Bought Leazes is a meadow with a sheepfold. The Havers grew oats. One field is named Part of the Ley called Brownheads. Wild Cat Crook names an animal, and people are named in Jane's Haugh, Taylors Close, Rogers bush (bush being the 'whins' where the sheep shelter). Among the rest are The Bodles, The Bodles Meadow and Bodle Hole Quarter which, referring to holes in the ground, may have been where spirits or the bogeyman lived. A Seggy Hole may come from sedge, especially the yellow iris, but it is more likely to be a source of clay to make saggars. Turby Crook is probably the ME *turbarye*, a place where peat or turf could be dug. A 'turbary' is a peat-drying place. The Maer may be from *(ge)maer*, a boundary. Spears may be from OE *spearcas*, covered with brushwood. Whip Hill may be from whip grass or a source of stems such as whitebeam or wayfaring tree, for making whips.

That is a considerable amount of information from one map. The maps that follow give much similar information and the main characteristics are summarised below each map, using a similar classification of elements.

ALPHABETICAL LIST OF COMMON ELEMENTS (DENOMINATIVES) IN FIELD NAMES

Examples of these will be found in the general index; the purpose of this list is to provide a quick reference system.

Acre	OE *aecer* = a plot of cultivated land which in the open-field system was a strip. In modern use it is a unit of area
Allotment	ME a piece of land claimed from waste or forest and ploughed
Assart	land allotted to a person or place
Baulk, balk	OE *balca* = unploughed division between units of land
Bank	ME *banke* = bank, slope of a hill
Bog	ME *bog* = marsh
Brae	ON *bra* = brow, edge of a hill, slope
Brake	OE *braec* = thicket
Broad	OE *brad* = broad; OE *braedu* = broad strip
Brook	OE *broc* = stream
Burn	OE *burna* = stream
Butt	ME *butte* = strip of land shorter than others in the same furlong; OE *butt* = tree stump; ME butt = mound, site for archery practice
Carr	OE *carr* = rock; ON *kjarr* = brushwood; ME *ker* = swamp, marsh
Cleugh	OE *cloh* = deep valley
Cliff	OE *clif* = scarp, steep slope
Cot, coat	OE, ME enclosure. OE *cot* = cottage, animal house. It seems also to mean 'lands on the border' in local examples

Croft	OE a small enclosure, usually next to a cottage
Dale	OE *dal* = share of the common land; OE *dael* = valley. Deal, dale and dole all mean 'divide' (H)
Darg, dorg	a day's work
Dean, dene	OE *denu* = valley
Dike, dyke	OE *dic* = a ditch, drain; locally, a wall. This frequently appears as a boundary, with a quickset hedge planted on the up-cast
Field	OE *feld* = (a) stretch of open country (b) open common field (c) enclosed plot of land
Flat	ME = a piece of level ground. Flatts were also divisions of the common field
Fold	OE *fald* = small enclosure
Furlong	OE *furlang* = the largest divisions of a common field, where groups of furlongs made up a 'great field'. They are also called shotts, sheths and furshots. Originally a 'furrow' length
Gair	grassy spot
Garth	ON an enclosure
Gate	(a) a stint, or right of pasture (b) ON *gata* = road (c) a gate
Gore	OE *gara* = triangular piece of ground
Green	OE *grene* = grassy;
Ground	OE *grund* = usually a large piece of ground
Hagg	OE *haga* = fence, fenced enclosure
Haining	OE (*ge*) *haeg*, *haegen* = enclosed land, meadow
Hale, Hall	OE *halh* = nook, land in the bend of a river, like 'haugh'
Ham(m)	OE *hamm* = an enclosure
Hanger, Anger	OE *hangar* = a wooded hill; OE *hangende* = overhanging, steep
Haugh	OE *halh* = originally alluvial land in a river bend, it came to mean any land by a stream or river
Havers	lands on which oats are grown
Hay	like 'haining', or ME 'hay', it is a forest fenced off for hunting
Headland	OE *heafod* = place where the plough turns, end of a ridge, source of a river, upper end
Heath	OE *haeth*
Hern	OE *hyrne* = a nook or corner
Heugh	OE *hoh* = heel, projecting ridge. In dialect it has become a crag, cliff, precipice. In general its meaning can vary from a slight rise to a steep ridge
Hield	OE *helde* = a slope
Hoe	OE *hoh* = see 'Heugh'
Hole	OE *hol* = hole, hollow, pit
Holm	ON *holmr* = water meadow, riverside land, dry land in a fen
Holt, Hot	OE *holt* = a wood, thicket
Hop(e)	OE *hop* = small enclosed valley, a branch valley, a blind valley. It can be dry land in a fen

Hoppet	OE a small enclosure
How	OE *hoh* = see 'Heugh' or ON *haugr* = natural or artificial mound. It can mean a hollow or depression
Hirst	OE *hyrst* = copse, wood, wooded hill
Hyrne	see 'hern'
Ing	ON *eng* = meadow, pasture
Inhook	ME *inhoke* = land temporarily enclosed for cultivation while the rest stays fallow
Inland	ME land near to the homestead
Intake	ON *intak* = a piece of land taken in from waste
Knoll	OE *cnoll* = hillock, hilltop
Lake	OE *lacu* = stream, watercourse
Land	OE a strip of ploughland in the common field
Lane, loan, lonning	OE *laning* = lane, track
Lea, lee	OE *leah* = wood, clearing, meadow
Leases, leazes, leys	OE *laes* = pasture, meadow
Letch	long narrow swamp in which the water moves slowly
Link	OE *hlinc* = undulating sandy ground; terraced ploughing; unploughed strip
Low	OE *hlaw* = mound, hill, burial mound
Mains	ME *main, mesne* = demesne land
Marsh	OE *mersc* = fen, bog
Meadow	OE *maed, maedwe* = grass land kept for mowing
Mire, mier, mere	ON *myrr* = bog, swamp, boundary
Moor	OE *mor* = moor, bog, waste
Moss	OE *mos* = Marshy ground where moss grows
Nook	ME *nok* = nook, secluded ground
Orchard	OE *orceard, ort-geard* = fruit garden
Over	OE *ofer* = slope, ridge, hill
Park	ME an enclosure for hunting or pleasure
Pasture	ME grazing land
Pen	OE *penn* = pen, fold
Pickle	ME *pightel* = small enclosure
Piece	ME *pece* = an allotment
Pike	OE *pic* = pointed land
Pingle	ME same as 'pightel'
Plantation	nursery
Plashet	OF *plaschiet* = waterlogged ground
Plat, plot	ME *plat* = small plot of ground
Pot	ME *potte* = hole, pit
Quarter	ME *quartier* = portion, allotment
Ray	ON *vra* = land in a nook or corner; ON *ra* = boundary
Riding	OE *rydding* = cleared land, clearing

Reen, rein	ON *reinn* = land on a boundary. It is also a terraced strip of land on a steep hillside showing signs of having been under tillage in the past, like a lynchet
Ridge, rigg, rig	OE *hrycg* = a strip of land in the open field, the basic unit of tenantry
Rood	OE *rod* = about 1/4 acre
Scribe	a long narrow strip of arable land, which at Warkworth averaged over 1/5 acre
Shaw	OE *sceaga* = small wood, copse
Shield	ME *schele* = hut, shed
Shot(t)	OE *sceat* = the same as furlong or flatt, a block of land in which all the strips ran in the same direction
Slade, slate	OE *slaed* = valley, low-lying marsh
Slough	OE *sloh* = boggy land
Spring	ME a copse of young trees
Stobbs	OE *stubb* = stumps of trees
Strother	OE land overgrown with brushwood
Waste	unimproved common pasture; a source of turf for fuel
Wash	OE *waesce* = sheepwash; OE *waesse* = wet place, swamp
Well	OE *well, wiell, waell* = well, spring, stream
Whin	gorse
Yard	OE *geard* = land by a building, enclosure
Yield	OH *helde* = a slope

A CONCENTRATED AREA OF STUDY: THIRSTON, FELTON AND ESHOTT

There are thousands of names available for study and comment, spread over many parts of Northumberland. In many ways the choice has been made for me by the availability of material. It would be wrong not to show the reader the depth to which the study can go, so I begin with a detailed examination of one region first, then look more broadly at other maps of other areas to discover what they tell us.

First, a group of maps from the same area. This is focused on Felton in the west, and to the east and north-east the area extends to the North Sea. In it is some prime arable land, some owned by the Duke of Northumberland. Felton, at an important crossing place to the north of the River Coquet, was named after open fields. Across the water is the township of Thirston. Together, they carry the route of the old A1 road, but are now bypassed. The medieval bridge, like the one at Warkworth, and that at Rothbury, emphasises the importance of such crossing places. Because Thirston has belonged to the earls and dukes of Northumberland, the lands are particularly well documented.

The river, flowing east along an often-precipitous valley, attracts names that reflect this fact. There are large numbers of haughs and banks, for example, and vegetation growing by the river such as alders, willows and elders. A document of 1250 records:

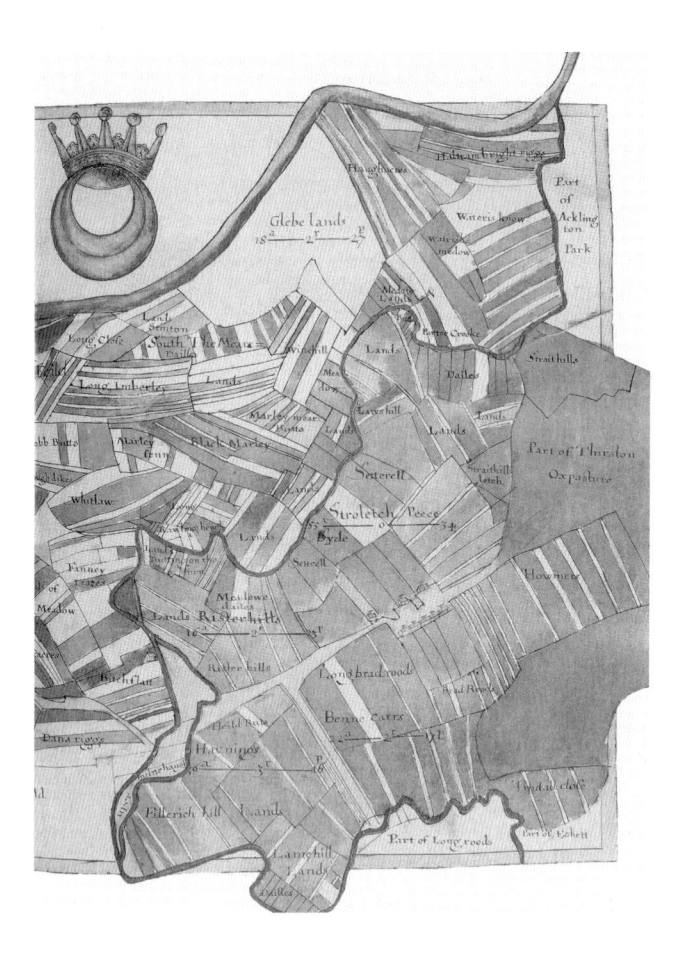

16 Thirston in the 1620s

A moor near the Chynerig in Thresterton field that was granted to Newminster Abbey. At about the same time 12 acres in Thrasterton were granted to maintain the light of the blessed Mary in Brinkburn church: 1 acre in Mickledayle, 1 acre in Langelands, ½ acre in the same culture, 1 acre in Ebrockes, ½ at Mosvcrokes, 1 acre at Crysedale, 1 acre at Biglichirne, 1 acre at Beneacres, ½ acre at Fulton (Felton), called Heuedes, ½ acre at Hendacre, 1 acre at Colsaw, 1 rood at Benefordacres, 1 rood at Wetenhalghford.

When we look at this, some of these names are still possibly in use: Langelands (Long Field), the Leys (Birk Lases), Ebrocks (Brock letch) and Beneacres (1772 Benacres, 1851 Binnacle Close).

The rest have elements that survive in many field names. Mikeledayle is land in the big field, lands and acres being units, Crokes are bends, and Mosscrooks would be boggy. The fording place at the haugh would be one of many, and *hirne* is land on the bend of a river. Benacres grew beans; Biglichirne grew barley. Benefordacres were bean-growing areas at the ford.

In 1257 William Puffyn gave eighteen acres to the prior and convent in the field of Thrasterton next to the culture called Monhalme. In 1772 the names Well Monaye and West Monday appear on the map. Moneye was land subject to a special money payment. In the 1330s, much of the land at Thrastereston was derelict 'because they lay utterly waste and uncultivated through the poverty of the neighbourhood destroyed by the Scots and lack of tenants and animals.'

In 1567 a very detailed survey was made for the seventh Earl of Northumberland. All the land was in 'husbandlands', and among the tenants were David Daye, William Garrett and Thomas Hudson. The names of tenants often named fields: there is a Days Peece in 1772 and a Garretts Head Lands. Hudson gave his name to a close in Felton. The vicar of Felton had 'le Vycars Halghs'. In 1772 it is called Rond Haugh Glebe and Havers Glebe; in 1851 it is Felton Glebe lands and Beneacres appears again.

In 1569 a pasture called Streyght-Hills is named along with Glovers Meadow (later Glover Close) and Gresse-yarde. The Stryght Hills appear again in 1855 and the site is marked on the 1623 map (by 1851 it is a part of Thirston Grounds). It could be the OE *steort*, a long projection, a tail of land.

This 1585 Survey, full of interesting detail, has a section that defines the 'Bounder', or boundary of the township. This serves as a good example for others in the County:

> Beginning at Elstrother yate and so Wintrick house, and from thence to Fertles dyke, and up that dyke to the Black bushes, and so up to the northe east end of the loning of Bockingfield, and from thence up the moore along the yeard to Gybbes Eshe, and from thence downe along by Pygles Close, to Burgeam yate, and from that yate to Selbyes foard, and from thence along by the foxeholes to Headlawe, and so to Headlaw wood head, and from thence downe the march bourne to the Water of Cocket, and down that water to a tenement of Mr Lyles, called the Catt-heughe doores, and so down along a dyke to the foote of Howdens at the Warter of Cockett, and so along the water of Cockett to Elyhaugh foard, and from thence to Shothaughe foard, and so to the Foggyleas steele, and from the Foggyleas style, along down Thryston dyke to Monhow born head, and so down along the Hamelspeth, and from thence up the dyke there to the foule-brigges, and from thence along Whormesley dyke to Elstrother yate, where we began.
>
> The inhabitants of Felton, Bockingfield, and Burgeam have entercomon by byte of mouth within the said bounder, as his lordship's tenauntes of Thirston have likewise within their bounders, and the said inhabitants of Felton, Backenfield, and Burgeam, as likewise his lordship's tenauntes of Thirston, do use to dygge, grave, and gette peates, flagges or turves within the same bounder.

In the North, a 'dyke' is usually a wall. The name Burgham probably refers to burgages, Catheugh is cat cliff, Howdens is a hollow valley, Elyhaugh is a small island in the river, and Shothaugh refers to ploughed land by the river (a furshott).

The Foggylease Steele (coarse grassland by a stile or steep place), Gybbes Eshe (Gibb's ash tree), and Pygles Close (ME *pightel* is a close, a small enclosure) do not appear later, but Monhow born (burn) may be linked to Well Monay, and Hamelspath is still called that today. The slope down to Felton bridge is still called The Peth.

In the list of tenants of 1585 John Tynedale and Thomas appear, so Tindales Close of 1772 (Tindalls in 1851) is explained. By now Gresse-yard had become Gryce Yearde. Thirston land lay open, and there was a heated dispute between Thirston and Bockenfield

88

in 1591 over rights to 'certen grounde'. William Fenwick complained to the earl that John Heron of Bokenfeeld abused the earl's tenants by 'beatinge them and strikinge and sleainge their cattell with lance staves and hounding them with mastiff dogs.'

We see in 1620 that 'the lord and tenants of this manor doe entercomon in the common called Bockenfeild moore and Elstrother' with the townships of Felton, Bockenfield, Weldon and Nether Framlington. The same document sets out the names of tenants, the area of their house and land and how much pasture they were entitled to, the area of house or cottage with garth, the amount of arable, meadow and the number of 'gates' (where 12 gates = 21 acres, 18 perches).

Much information about fields, therefore, can come from documents, and it now remains to look closely at the maps, of which there are three main ones. The information on Robert Norton's map of 1623 and meanings may be classified as:

Physical features
Watrish medow, Wateris Know, Straithill (OE *sterot*, a projection), Laws Hill (OE *hlaw*, hill), Snabb butts (projecting hill), Heild buts (sloping land), Marley fenn, Well hill (a spring), Whitlaw (white hill), Saugh butts (willow), Fanney Leazes (fern covered meadow), Bitch-flatt (birch), Stroletch Peece and Straithill Letch (slow moving watercourses).

Position of the fields
Southeran hills, Above ye milne, West Crofts, Farre Close, Endicarr, Land sand butts on the Burn, Upper Milnehaugh and Lowe field.

Shape and size of fields
Long Imberley, Long Close, Part of Long Roods, Long brad roods, Lang Howden (valley), Brad Meadow (broad), Brad roods, short acres, Short Riggs.

Land divisions
These follow the pattern of the open-field system: Acres, Haughacres, Dailles, Dana Riggs, Haltram bright riggs. A flat of land.

Agriculture
Haynings (enclosure), Benne Carrs (beans), Lands Rister Hills (rye), Lamehill lands (ME *leme*, the land had to be drained artificially).

Others
Hending riggs (arable strips on the hen pasture), Mary Meadows (possibly a religious dedication), Halleywell leazes (holy well), The Mearc (the boundary at the Glebe lands); Marley Mearc Butts is also a boundary, Colchas (possibly Col-Shaw, charcoal burners' wood) and there is some fine clay in the area. Arbour lodge is either an arbour or OE *eorth-burg*, an earthwork.

These have to be seen on the map in relation to strips of land of different kinds of holding.

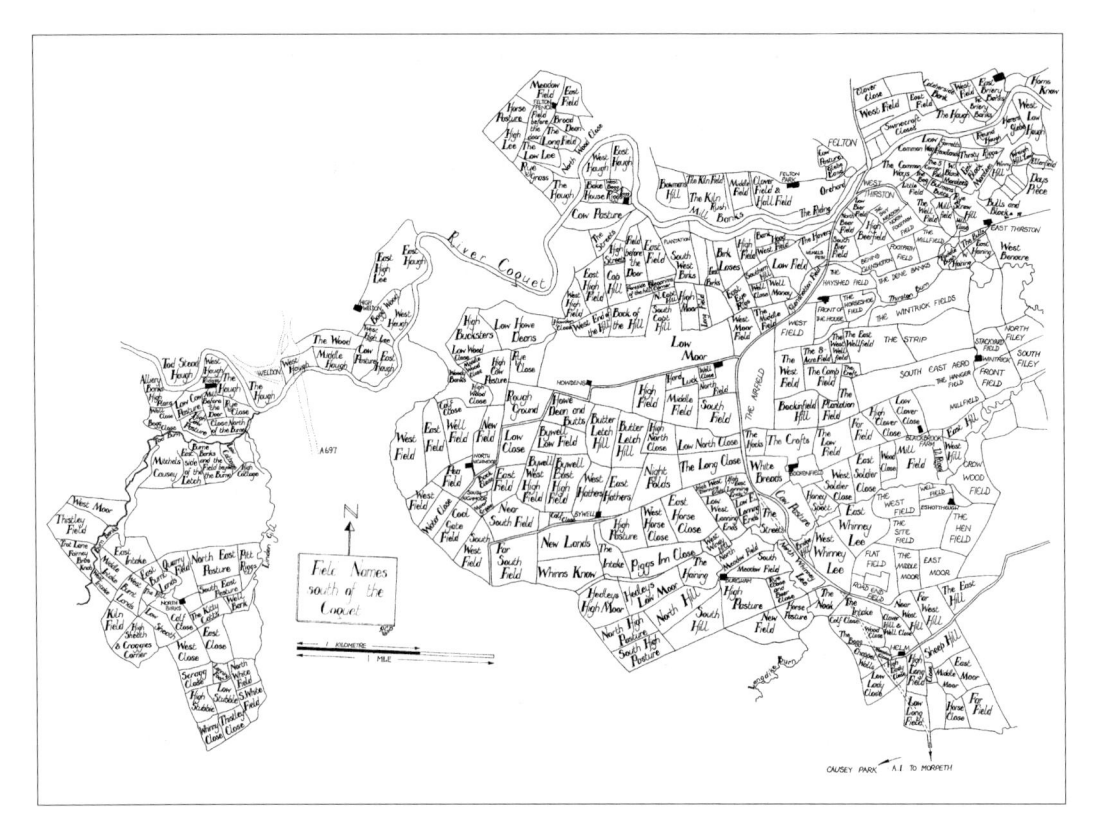

17 Modern fields south of the Coquet

Comparison: 1772 and 1881

Taking about 100 years between maps provides an interesting reflection on the changes between those dates. In most cases there has been no change in the size of individual fields since 1772. The name given first is 1772, with 1851 names bracketed where there has been a change.

Low Common Way (2 unnamed fields)
High Common Way (The Common Ways)
Garretts Headland, Bog Close
Thirsley Rigs (Thistley Rigs)
Unthank Close
Round Haugh Glebe and Havers Glebe (Felton Glebe Lands)
West Low Haugh and East Haugh (East Low Haugh)
West Haugh, Winny Hill (not named)
East Black Moralee (High Moralees and Low Moralees)
West Black Moralees (Harbottle's Field)
Rye S…Hill and West Half Crown (Rester Hill), Butts and Black (the Calf Close)
Wreigh Hill Law and Ellenside Bank (some individual plots)
(The Rye Hills)
Days Pece (The Days Piece)
Wreiglaw Hill (two plots)
North Bearfield (Low Bier Field)
High Beerfield (High Bier Field)
Mr Redhead's land (Hemmel Pathfield)
not named (Charley's Field)
The Havers (not named)
Bankhead and West Field (The Wood Field)
Low Field (Harbottle's Field)

Well Monay and Well Close (The Southern Field)
West Monday (not named)
Southern Hill (Corby's Hill)
High Field (Raven's Crag Close)
Birk Lases (East Coquet Bank)
East Birks and Middle Birks (Middle Coquet Bank)
East Eye Rig and Mr Readhead's land (East Sunniside)
Long Field (Middle Sunniside)
North Shothaugh Lane, and North West Birk (Plantation)
South Shothaugh Lane and South West Birks and High Moor (West Coquet Bank)
North Copt Hill and South Copt Hill (Copt Hill)
unnamed land (West Sunnyside)
Low Moor (The Moor Field)
High Field (The West Field)
Hard Luck and Well Close (The Well Close)
Middle Field, North Field (Front Field)
South Field, West Moor Field (The West Field, The Eight Acre Field and The Camp Field)
Bockenfield Hill (The Plantation Field and Bockinfield Hill)
East Moor Field (other owners)
Far Horners (East Horners)
Hither Honers (West Horners)
Brock Letch (The Brock Letch Close)
Sandy Close and Calf Close (Sandy Close)
Tindales Close (Tindalls Close)
High, West and Low Benacres (Binnacle Close)
Toddle Bush (Toddle Bushfield)
West Haining and East Haining (The West Field)
The Haugh and Clover Close (The Horse Close)
High Rife (Cheeveley Head)
Low Rift (Cheeveley Door)
West Rift (The Saughy Pond Close)
East Rift (The Back of the Hill)
West Rigs Hill (The Near Horse Close)
The Shaw (The Little Wood Field and The Big Wood Field)
South White Faugh and Middle White Faugh (The White Fallow)
Lidget Moor (Sedges Moor)
South White Faugh (The Low White Fallow)
High White Faugh (The High White Fallow)
North Low Field and South Low Field (Plane Tree Field)
The Galls (The Hill)
East Half Crown (The Big Half Crown)
Cow Lone (The Cow Loan)
Little Close (The Narrow Close)
South Field (The Front Close)
Whinny Close (The Big Field)
Fryar Pool (The Moor Close)
Gurls Gap, North Howdeen
South Howdeen (Thirston Grounds belonging to Andrew Robert Fenwick Esq.)
Mr Gabriel's Freehold (The Five Corner Field, The Bog, The Well Field, Bulmans Butts, Short Acres, The Mill Field)

Such a list is not easy to take in, but it does show how field names change, and which elements remain constant. It also helps to confirm meanings.

These data are now arranged to show what the names tell us:

Physical features
Copt Hill (OE *copp* = hilltop), Southern Hill, The Hill, The Whinny Hill (Whin is furze, gorse, but OE *winn* = pasture), Bankhead, Middle, West and East Coquet Bank, Howdeen and Howden (a hollow or deep valley), The Haugh (and all the locations of these by water), The Brock Letch (brook moving slowly through a narrow swamp), Bog Close,

The Bog, Black Moralees (swamp clearing), Sandy Close, Far and Hither Horners (corner – in fact enclosed by a sharp road-bend), The Moor Field (waste), The Wood Field, Sedges Moor, Ellenside Bank (OE *elren* = elder trees), The Shaw (a wood), Plane Tree Field, Saughy Pond Close (willows), Whinny Close, East Birks (birch trees), Clover Close.

Position
North, south, east, west, high, low, middle, front, far and near, the back of the hill. It may lie near to a named farm or village (Cheveley Head) or some other feature (The Well Close, Mill Close). There is also a Sunnyside.

Size and shape
The Little Close, Short Acres, The Narrow Close, The Big Field, The Five Corner Close, The Eight Acre Field. There is a 'Big', 'East' and 'West Half Crown', a name which usually refers to its size and shape, but in the 1680 Felton Register it says, 'It was agreed by the minister and those of the four and twenty present, with the church wardens, that a sesse of halfe a crown be laid upon every plough in Felton parish for the paving of the path and mending the high waies to prevent the parish being fined.'

Agricultural practice
Terms, such as acre, butts, land and dale were specific measurements. There are also many indications of what was grown: Birk Lases (leas = pasture), The Havers and Havers Glebe (havers = oats), Bierfields (and other spellings, (OE *bere* = barley, ON *bygg*), Beneacres (beans), East and West Haining (enclosed land, a meadow).

There is one group of fields where the 'White Faugh' is the common element. Later the name was changed to 'Fallow'. White is the light colour of some grass. The beginning of new tillage appears in High, Low, East and West Rift. Farm animals are named in Cow Loan, Calf Close and the Horse Close.

Non-domestic animals
Toddle Bush (fox hole or fox hill), Corby's Hill (crow), Raven's Crag.

People's names
Harbottle, Readhead, Garrett, Tindale, Charley. In Bulman's Butts the name could be his occupation or surname. Moralees is a local surname, but here could be a swamp clearing bordering the Coquet (OE *moriga(n) – leag*).

Sounds
Spelling is not usually important in oral tradition. Wreigh Hill in 1772 becomes Rye Hill in 1852. Near Rothbury, Wreigh Burn is pronounced 'rye' and comes from OE *wearg* = a felon, but in the same area Wreighill is pronounced 'ree'.

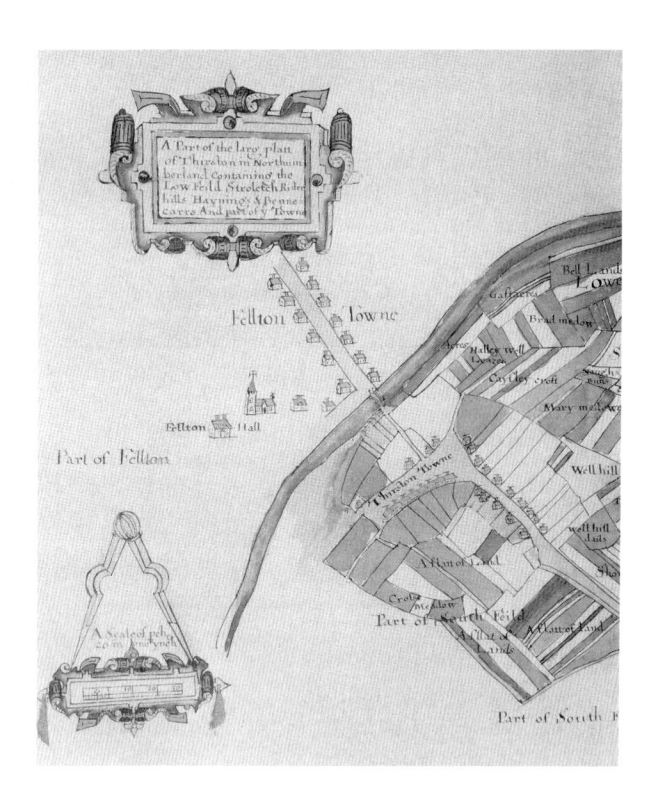

18 Felton Tonwship in the 1620s

Social
Unthank Close is a squatter's enclosure, Felton Glebe Land is owned by the church, the High and Low Common Way are fields where there is a public right of way. 'Hard Luck' speaks for itself. Well Monday, Well Monay and possibly Eye Rig could be 'Moneye', or land on which a special payment had to be made.

Others
Hink Close and Gurls Gap? The Galls can be accounted for by 'oak-galls', swellings on trees caused by insects. Oak gall is used in the manufacture of ink, tannin, dyes and medicines. Could Fryar Pool be for fish fry, or have some connection with a friar?

Felton township
Only the river separates Felton from Thirston, but the sources of field names are different. In the 1770s Thomas Bell produced detailed maps for both areas, and the Tithe maps of the 1840s are similarly detailed. A summary follows.

The River Coquet and its tributaries account for the large number of haughs, banks, and names such as Broad Dean, North Dean, Low Howe Dean, Howdon Close (valleys) and Water Close. Although there is no dramatic rise and fall in the land except at the river in parts, there is sufficient to account for the number of 'hills' and 'cop', which means the same thing. Know is a small rounded hill, as in Whins Know and Fairy Knows. The Snook is a projecting ridge.

Bog Field, East How Mires, Moralees and Rushy Close, Thistly Riggs, Rough Ground, Briery Bank, Broomy Field, The Scroggs (brushwood), Rye grass and Clover Field give a

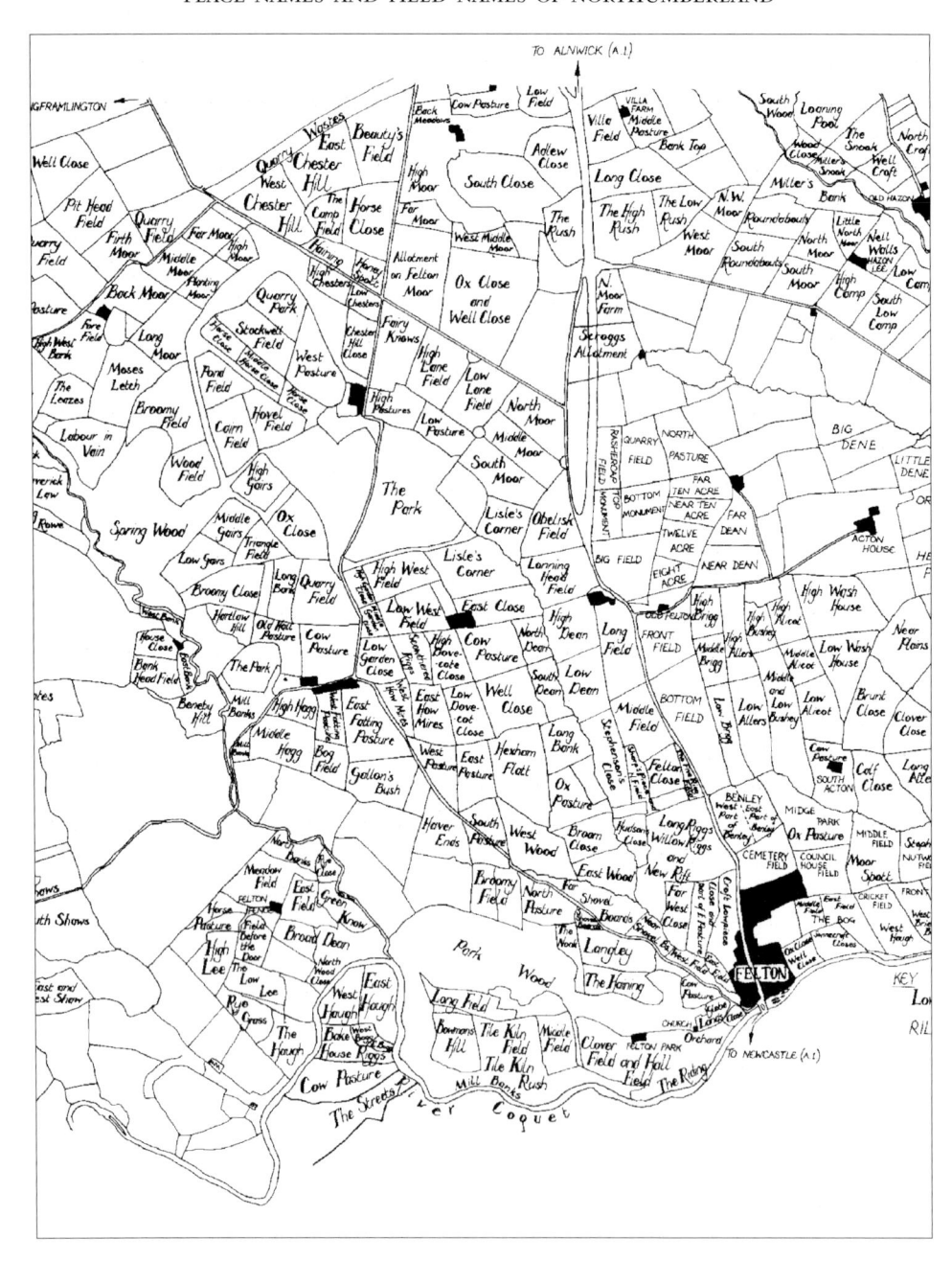

19 Fields north of the Coquet

picture of soil conditions and vegetation in those fields, and in the woodland areas Low Allers and Alicot (alders), Ellenfield (elders) and Willow Riggs are more precise than North Wood Close or South Wood. High, Middle and Low Gairs are strips of green once surrounded by scrub or heather. The shape and sizes of fields are given in names such as Triangle Field, Roundabouts, Long Field and Long Riggs. The Nook is a small field. Shovel Boards means as broad as a shovel – a fairly common local name for a narrow strip of land. White Bread is a broad field with whitened grass. Many give the position in relation to others. Some, such as slow lane Field, High West Lonning Ends (lane), Footpath Field, High lane Field, Bridge Hill, Hall Field, Well Croft, Obelisk Field (a memorial to Nelson), Bake House Riggs and Dovecot Close take their names from nearby man-made features, and so too do those fields connected with industries – Quarry Field, Pit Heap Field, Tile Kiln Field and Kiln Close.

There are still many examples of rig and furrow to be seen, some going back to the open-field strip system of farming, and there are reminders of this in Long Riggs, Bake House Riggs, Scowthered Riggs, High Brigg, the Five Rigg Field, The Butts and The Dales. Oxen, horses, swine, cattle and sheep have their associated closes. Night Folds are where they were penned at night. Meadows and pastures are well represented and Fatting Pasture and Stockwell Field are where the animals grazed. High Wash House is possibly where sheep were dipped. Parks were reserved for game birds and animals. 'Haining' is an enclosure or grove, often reserved for hunting. Hagg is also an enclosure.

From time to time new ground was brought into cultivation, often from common or waste land. 'The Hacks' may be from ME *hacket*, a piece of cleared ground, or from OE *haca*, a thorn tree. Intake, New Rift, New Field and New Ground record this opening up of new land. Rye Straw Hill, Rye Close and various names that mean 'barley' (Beer Close, High Beerfield, South Bierfield), Haver Ends (oats) and Wheat Field tell of grain being grown. Pea Field and various bean fields (Beneby, Benley, West Benacre) grew pulses. Butter Letch Pasture is good grazing; Labour in Vain and Hard Luck the opposite. Honey Spott is the place where beehives were kept.

Chesterhill is a pre-Roman earthwork, and further east there is a 'camp' element in three fields. Although there is no longer any trace of burial mounds, 'Cairn Field' in Swarland Wood used to contain at least three. Care has to be taken with names such as 'High Streets', which in some contexts can be a Roman road; here it is from OE *steort*, a projecting piece of land.

People's names remain in Lisle's Corner, Stephenson's Closes, Sharp's Piece, Gallon's Bush, Bowman's Hill, Bulman's Butts, Hedley's High Moor, Days Peece and Hudson's Close. There are problems with some names. For example, is Beauty's Field named after a horse, a person or neither? Similarly, why West and East Soldier Close, Glegarims Corner and Nell's Walls?

Eshott Estate
Eshott is named after a ridge (heugh) to the south of Thirston, on which ash trees grew. A flat area leads up to it, used as an airfield, with a small operation for microlights and other small planes today. The ridge is one of many in Northumberland that commands wide views without being very high, in this case both to the north and south. Nearby Helm may refer to the helmet shape of the ridge or to a roofed shed. Eshottheugh incorporates OE *hoh*, a projecting ridge, but hardly a cliff.

The two maps are based on a sale catalogue of 1877 and one made in 1976, discussed with me by the owners, Mr and Mrs Sanderson. I have already referred to the naming of three of these fields by them: High Seas, March Brown and Quality Walk. There is great

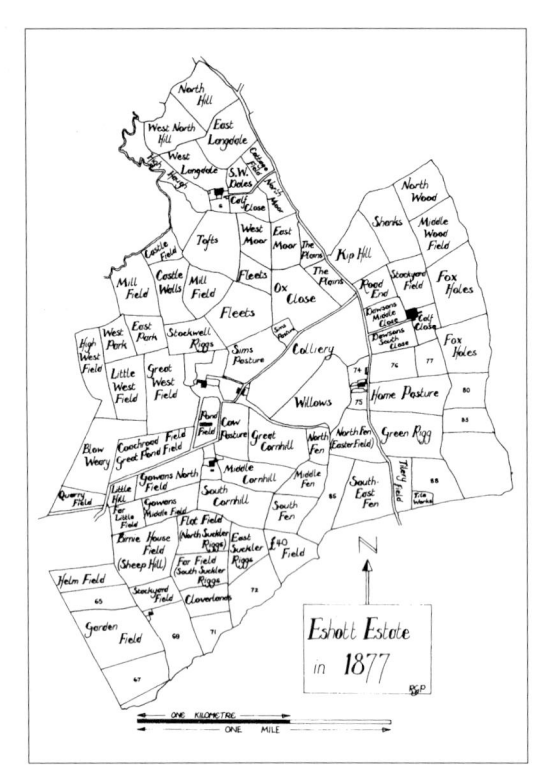

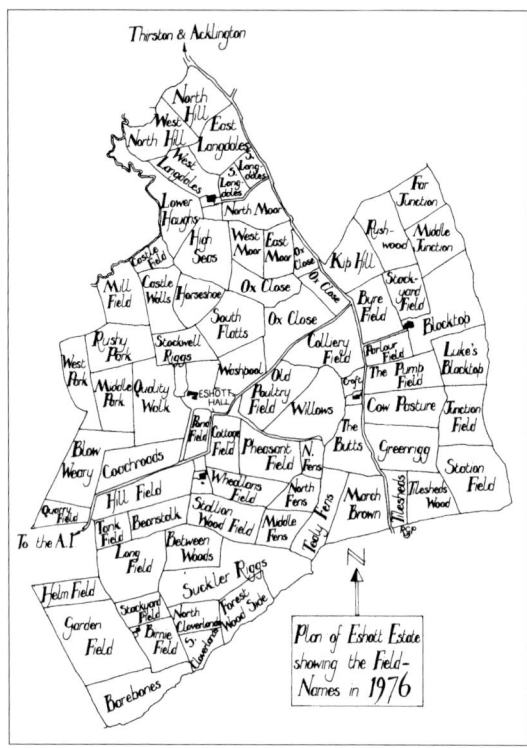

20 Eshott in 1877

21 Eshott in 1976

continuity in the choice of older survivals, such as tofts, dales, butts and riggs. People's names are included (Dawson, Sims, Gowan, Luke, Wheallans, Brown). The function of the fields – Stackyard, Colliery, Mill, Quarry, Corn hill, Cloverlands, Tilery, Byre, Poultry, Stockwell, Beanstalk, Garden, Washpool, Pump, Cow Pasture, Ox Close, Stallion Wood Field and Cricket are defined. Eshott Castle, now only an earthwork, still gives its name to Castle Fields and Castle Walls. Fleets are streams. Fields are positioned by the end of a road, a coachroad, a station and junction, their relationship to Eshott Hall, a pond and a cottage. Blow Weary (Blawearie elsewhere) speaks for itself, Barebones does not sound like good land, nor do Fens. There are rushes, willows, pheasants and foxes.

ACKLINGTON

I have always found the Acklington Tithe map of the mid-nineteenth century to be one of the most interesting. The same area is also covered by one of Robert Norton's maps, part of the Barony of Warkworth, the inheritance of the Earl of Northumberland. The latter shows the village built around its main street with small plots and cottages fronting on to it, with three open fields divided into smaller ones, the nearest to the settlement being the North, South and East and West Crofts. The cultivated land, distinct from the yellow coloured Common land (mainly 'Oxpasture'), has large fields which show signs of having been enclosed strips; lands, flats, acres and riggs form part of the names. Some of the names reappear in the Tithe map.

1 Bamburgh Castle: one of the oldest documented sites; much of what can be seen today is reconstructed

2 Chillingham: originally named after Ceofel, the settlement is now focused on an early castle and church, with a landscape setting of fields, woodland and gardens

3 Rothbury from Lordenshaw: the moorland, containing a wealth of prehistoric sites, gives way to cultivation in the Coquet valley. Lordenshaw may mean land that is difficult to cultivate on a ridge

4 Hadrian's Wall signpost east of Housesteads Roman fort

5 Harehope valley, Old Bewick Moor, from the west; a landscape that has no individual fields and remains 'waste'

Above: 6 Alnmouth: the estuary accounts for the small town as a harbour for grain. Beyond are enclosed fields of the coastal plain

Left: 7 South of Yeavering Bell: hill-farming country, with some arable in valleys and timber plantations. Yeavering was a prehistoric and Anglian capital

Right: 8 Belsay Castle: a medieval tower with extensions, part of a landscape that gets its name from a ridge called *Bileshoe* in 1162

Below: 9 Font Burn: now dammed as a reservoir; 'funt' in 1200 is a pre-Roman name, probably meaning spring

10 Alnham: a landscape of hills running into fertile fields on lower ground to the south

11 Holystone: the bend in the River Coquet encloses flat 'Haughs', with underlying gravel, surrounded by sandstone hills

12 Shortflatt Farm, near Bolam, with pronounced rig and furrow systems of the 'open-field' system of ploughing

13 Ingram valley: ancient field systems of lynchets, terraces and rig and furrow, the oldest of which are prehistoric

14 Hexham: beneath the Abbey is a Saxon crypt of the seventh century. The street plan reflects its ancient role as a market place as well as its importance as an ecclesiastical and civic centre

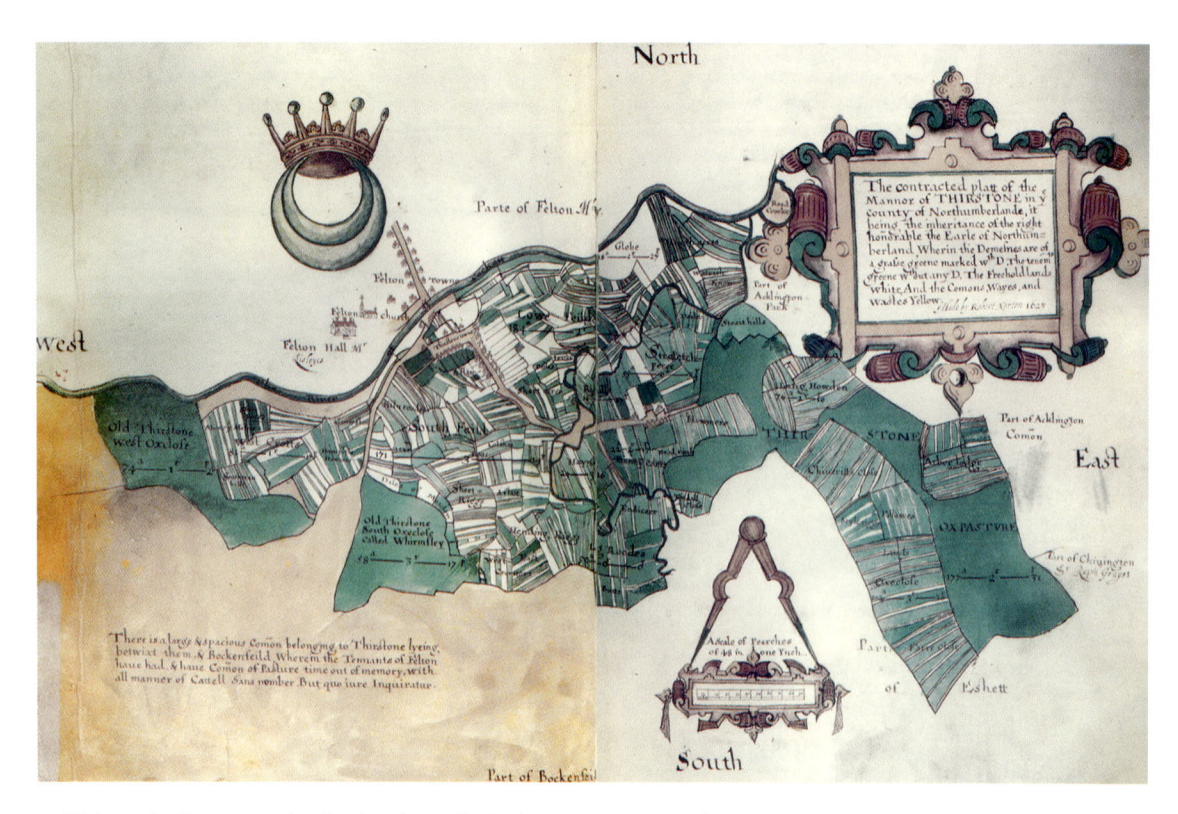

15 Thirston in the 1620s: a hand painted map by Robert Norton, one of many prepared for a survey of the Earl of Northumberland's estates

16 Felton from the east: its importance is marked by a ford and bridge over the Coquet. Its name means open or cleared fields

17 South from the River Coquet over Bockenfield to Eshott

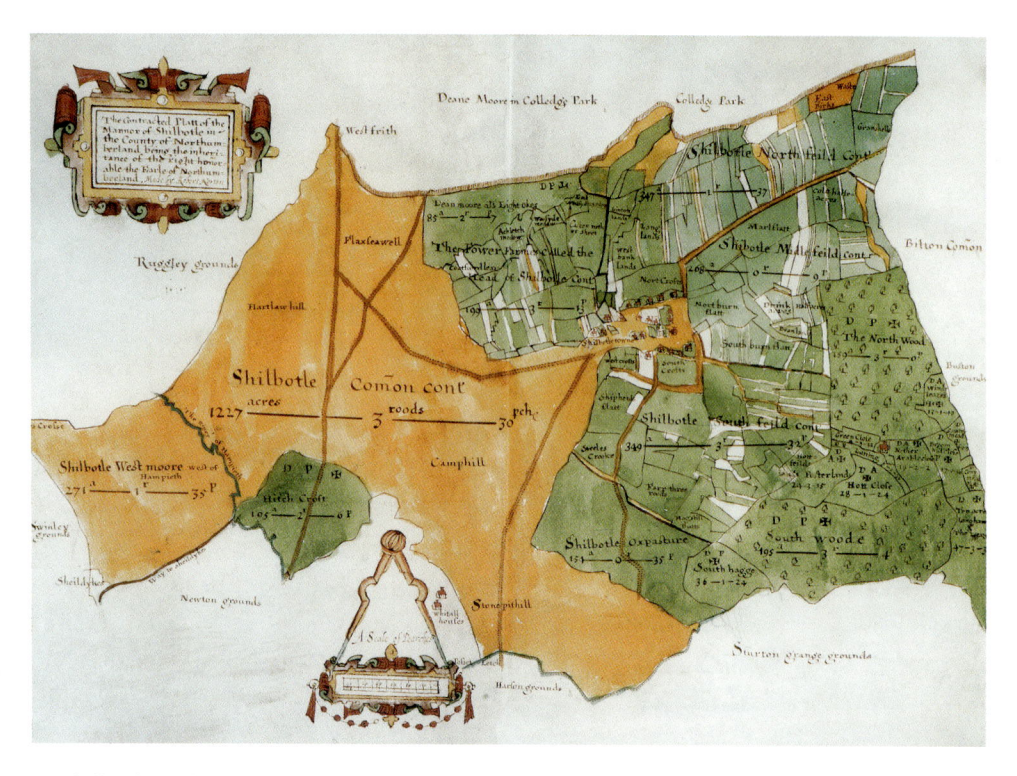

18 Shilbottle in the 1620s. The yellow areas represent common land; the others are strips in the big field with different kinds of tenancy

19 Lesbury in the 1620s: originally manor of the leech, or physician

20 Acklington, Warkworth and Buston in the 1620s. Standard symbols are used for houses that lie in plots of land on either side of the road

21 Bilton in the 1620s

22 Warkworth: one of the most important Percy holdings; the position of the castle and town in the bend of the river was originally Werce's enclosure

23 Warkworth: from the air, the pattern of surrounding fields and the core area are clear

24 Signposts like this have a deep-rooted meaning. *Courtesy of Matthew Hutchinson*

25 Holy Island, earlier called Lindisfarne, was central to the early spread of Christianity

26 Brunton to the sea: a typical pattern of enclosed fields, with sharp boundaries

27 Newbiggin-by-the Sea: although occupation can be traced to the Mesolithic period of prehistory, most of what we now see is modern

28 Ellingham: the field pattern east of the A1 road to the sea over Tyneley

29 Dilston: power base of the Jacobite Earl of Derwentwater, whose estates went to Greenwich Hospital after his execution

30 White fields: the possible explanation of this name may be seen in this seeded area of Fylingdales Moor, North Yorkshire, after a huge fire in 2005

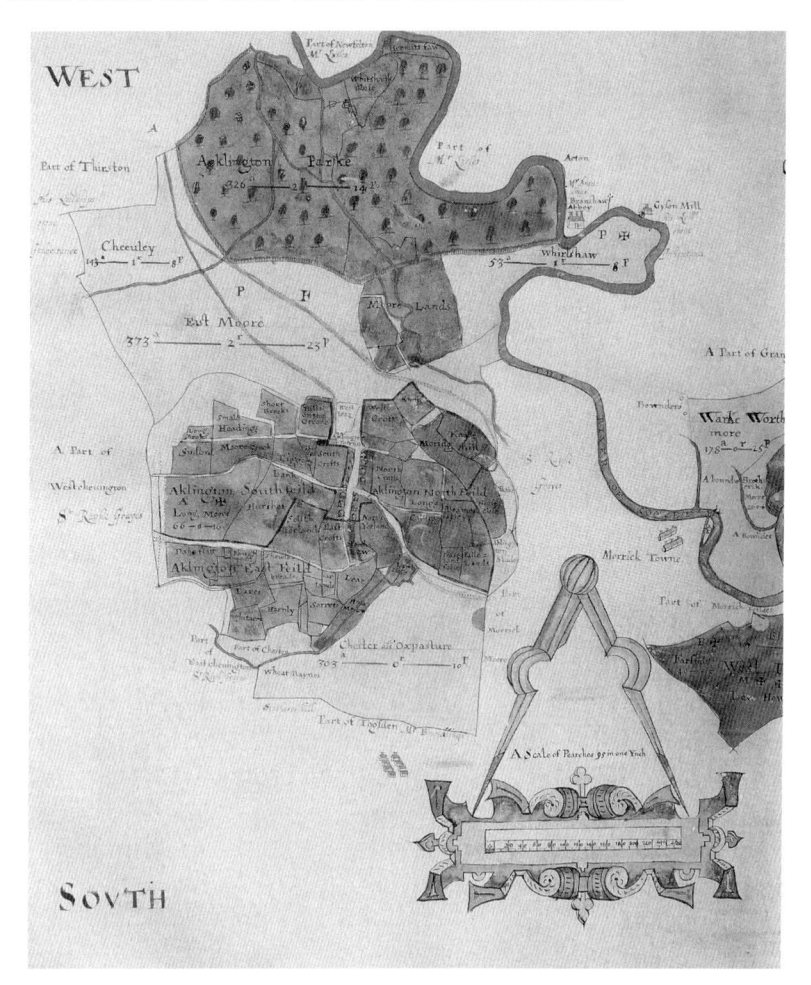

22 Acklington in the 1620s

What appears later as 'Hills of Grain' is named Hills on the Greene on Norton's maps. Moore Crook is there, Short Brocks, Long Moore and Small Headings (holdings?). Pasture is named as West Leaz. One field is named Harseshoe. The personal name Garrets goes with a field, and we have seen this name at Thirston. Kay Hill is where the cattle grazed, next to Moridge (an unproductive ridge). Carrlands may refer to land cleared of brushwood or stone. Reams may refer to a strip of land or a boundary strip. There is also a Wheat Raynes. Slades may be OE *slaed*, a low-lying valley. South Forland lies next to the Crofts and there is a Sudland (southland).

Acklington Park lies to the west of the settlement, mostly tree-covered, with Hermits Law overlooking the River Coquet. Just outside the park, encroaching on the common pasture, is a cultivated area still named Moore Lands. Most of this common pasture is East Moore, but one part is Cheeuley (Cheveley). To the east of the settlement it is called Chester Oxpasture, suggesting that there is an ancient earthwork, and a small piece called Wheat Raynes. The north side of the pasture includes Whirlshaw, Branshaw Abbey and Gyfon Mill (now Whirlyshaws, Brainshaugh and Guyzance).

Looking further back, in 1248 there were 21 bond tenants with 30 acres each. Only *Rumedu* appears as a field name. In 1352 there were 70 acres of demesne land and 7 acres of meadow; there were 35 bondage holdings, of which 9 lay waste. In 1472 there were still 35 holdings, with these names included: *Halle-stede* and a meadow called *Ermet-fall*. Twenty-six years later there were only 18 tenants, but they farmed all 35 holdings. There were eight cottage tenants. Two woods were named: Whorlecharle (Whirleyshaws) and Shevley (Cheveley).

A survey in 1567 showed that 'there is no comone land to be improved', despite the fact that there was a large common 'because of the barrenness thereof.' It recommends enclosure with:

> a strong quick hedge, and that the same so inclosed did lye twoo or thre years *in haninge*, in which tyme ye tenants mighte with there owne labour brynge ye same to a fyne grounde or at ye least to arable ground wher nowe yt is but *rotten mosse grounde*, which wolde be to the tenants in *grease tyme* much comodetie.

The words picked out in the text appear in field names. The same survey comments on the 'inequalitie' of the goodness of the ground, but does not advocate any change in the distribution of holdings; meanwhile 'acostomed lononge and common passadge' allowed peasants to reach their shares.

In 1585 these names appear: a croft called le Hole, a close called Green garth, a close called Howy's close and a meadow called Lambe meadow. The 1616 land survey is the most detailed; we see every piece of land measured, and there are details of whether it is meadow, pasture or arable. The common lands were large but 'somewhat barren' and would be better enclosed, such as the pasture ground called Whorlton Carre. Eighteen tenants held about 764 acres; 7 cottagers held 43; there were 1,169 acres of common pasture and waste; and Acklington Park was 714 acres.

A 1710 survey shows that, although there was the same number of farms, there were 8.5 on the north side and 9 on the south. The coal mine was working then.

This brings us to the detail of Bell's survey of 1839 and the Tithe maps. I have arranged the information to draw attention to physical features, position, size, agricultural practice, industry, personal names and names that pose problems.

Physical features
The Bank Head, Bank Head Field, Sandy Burn Fields, The North Haugh, Moor Crooks, Fairney Hill Banks, Thistley Moor, Fairney Hill (N, S, SW, SE), Soddy Brigg Close, Wood Close, The Wood Close, The Rush, The Long Brocks, Hoeings (North and South), Small Holdings, The White Moor.

Position
Moor Lands (First, Second, South), The Croft or Front Field, Lee Moor (NE, NW, S), Felton Field, Far South Moor, Felton Field Banks Wood, Near South Moor, Close (The Well (2), Near Well, Far Well Close, South, Corner, Fore, Woodside, North Moor (2), West and East Cheevely Close, Croft (Middle, East, West), Flat (Middle, Far, Near),

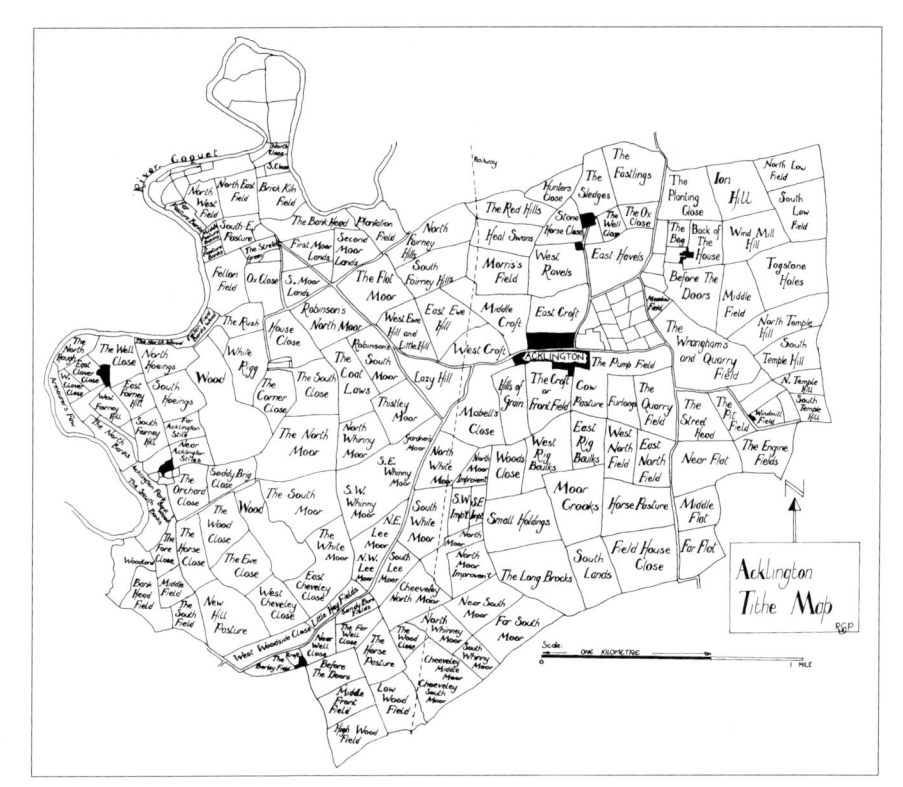

23 Acklington Tithe map

South lands, The South Banks, West and East North Field, Far and Near Acklington Stile, North and South Low Field, Acklington Park Bank Woods, Middle Front Field, Before the Doors, High and Low Wood Field, Back of the House, Field House Close.

Agriculture
Little Hay Fields, New Hill Pasture, The Barley Field, Middle, S. and Pasture Banks, East Havels, West Ravels (*havers* = oats), South East pasture, Hills of Grain, Meadow Field, The Orchard Close, The Stretch Green, The Ewe Close, Plantation Field, East and West Ewe Hill, Lazy Hill, The Ox Close (2), White Rigs, Cow Pasture, West Rig Baulks, Stone Horse Close, Furlongs, The Horse Close, The Horse Pasture (2), North Moor Improvement, North and South Improvement, The Fastlings.

Industries
Brick Kiln Field, Windmill Hill Field, Windmill Hill, The Quarry Fields, The Engine Fields, The Wrangham and Quarry Field, The Whimseys, The Pump Field, The Pit Field, Togstone Holes.

Personal names
Hunter's Close, Robinson's North Moor, Gardner's Moor, Morris's Field, Ion Hill, Mabell's Close, The Armourer's Field, Heal Swans.

Others
The Street Head (road or a projection), South and North Temple Hill, The Rey Hills, The Sledges.

The meaning of many of these names is fairly obvious; some of the more difficult ones are incorporated in the selective alphabetical list.

SHILBOTTLE AND BRAINSHAUGH

Shilbottle
The modern village of Shilbottle, with its spread of houses, lies to the east of the original centre, the remains of which are a green and a former vicarage called Peel House. At its east end this is a small medieval tower extended in 1863, with a barrel-vaulted first floor. The church belongs to the same recent period of building. The modern buildings were mainly devoted to the mining population, but the mines have closed, and new housing developments have recently been added.

The earliest documents include some field names:

1235   Westemestemede: the most westerly meadow, Shovelbred from ME *shovel brade*, a narrow piece of land, as wide as a shovel. Bene Flat is land good for growing beans, Blakesletch is black or bleak slow-moving water. Caldenelburne is a cold stream in a valley. Swinleys is swine pasture (later to be Shield-dykes), Remelde (later to become Rimside) from OE *helde*, a slope.

1472   Swynlees appears again, Baronhouse, Rymessid (rimside), Dowkerhalgh (ME *dowker* is a diver and *halh* is land by the river). Waterlees are water meadows, Tenacres are 10 strips of land, Tiallez later becomes Tyle-lea, a meadow near which tiles were made.

1567   Neither Shield-dyke was the far shieling ditch or wall where there would have been a temporary herd's hut. Barron's house, Barron's folde and Rymside appear again. There is Rugley lonying, (a rough pasture track), Cawledge park (crow's ledge or letch), Carter Dean meadow, Graindge borne was the grange burn, the outlying part of Brainshaugh abbey. Possetts Letche may refer to a drink, a posset. Espette fourde is from OE *aespe* – an aspen tree. Hampeth Fourde was the homestead-path ford. There was a Black close dyke and a Hedge-croft.

The 1567 survey of the Percy estates says that all of Shilbottle was in 'cottage and husbandland', the tenants having their land 'rigge by rigge to his neighbour according to the old devysion of lands in this countrye'. The township was 'a very poor towne', for although there was much arable land it was 'a waisted lean land for that they are not able to donge yt as the same wold be'; it had bare clay that needed manure and soil. However, there was good pasture – most of which remained unenclosed until about 1758. The Park was, in the Earl of Northumberland's accounts in 1472, called Shilbottle Park,

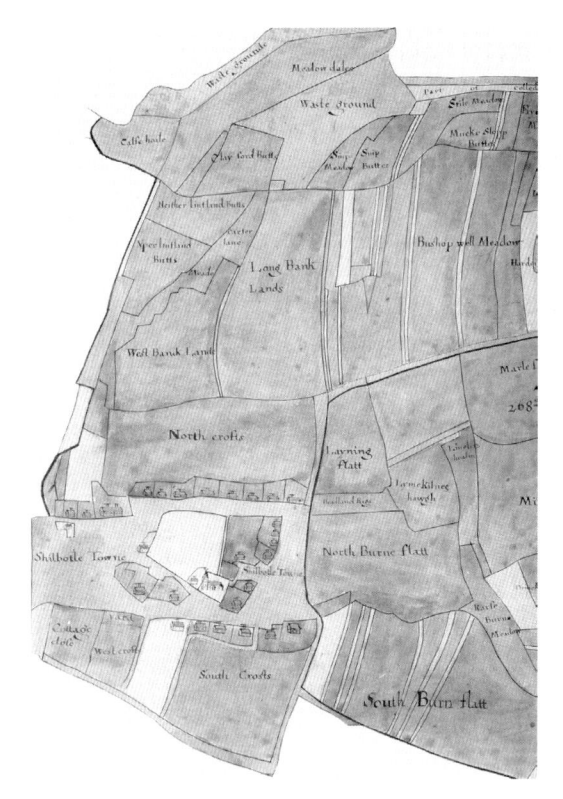

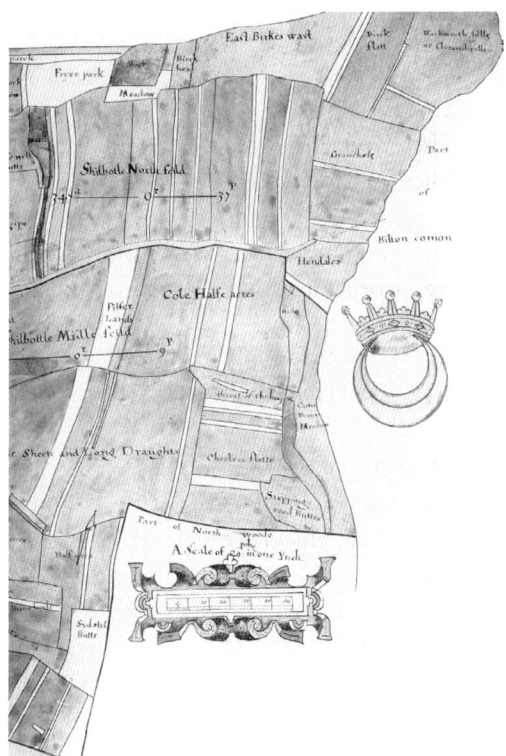

24 Shilbottle west, 1620s   25 Shilbottle east, 1620s

and was later known as Shilbottle Wood-House. The demesne meadows were called Dowkerhalgh, Waterlees, Tenacres and Tiallez (now Tylee Burn). In 1588 the closes were called the Holte, Forsterlaunde, Langhaughe, the Southe-wood, the Langhaugh-pece, the Over-seavenacre, the Under-wod-pece, Tyle-leae, the Style-hill, the Salter-meadowe, Cannon meadowe, Wanda-leaze and Carterdeane meadow.

The 1620 map names many of the open-field divisions – dales, butts, lands, riggs, acres, flats, roods; along with headlands, meadows, crofts and closes, Middle Sheete and Long Draughts are strips of land in 'parcels'. Some land is rough (Ruggley Grounds); there is clay, marl and waste. Some is covered with birch (East Birkes Waste), and there is plenty of woodland (West Frith, for example). Bushop Well Meadow is a spring in a bush-covered valley. There is a Carse Burn Meadow that could be named after watercress, or it could be from 'carr'.

Stonepithill, Lymekilnes hawgh and Cole halfe acres name industries which have continued ever since. Chesters flatte looks like an ancient fortification but Fryer Park Meadow could be influenced by Brainshaugh Priory and other religious foundations. Mucke Slopp Buttes is very evocative. Stepping rood Buttes is probably land ploughed originally in a clearing (OE *stybbing*). Sudslyl Butts are on the south side, Hitchcroft is a croft enclosed by hurdles (ME *hiche*). Granshels and Greanshields are shepherd's huts on pasture. A rather difficult name is Pilfer lands, which could incorporate pool (OE *pwll*) or come from ME *piled* (OE *aesc*), the bark of an ash tree that was peeled off to make a kind of sugar. The rest of the names are explained in the index.

101

Brainshaugh
Two sources of names are listed here:

1567   Whitt leases, The banke, Staynge leases (stony meadows), Morke Hawghe (Morwick), Braidle nowe parcel, Barnhyll, Newghe-dycke, Buke bushe peace, Binlee parcel, Kirpeswell hawghe, Midlewood, Marche-hagge, Lee closes, Moerlee, Haisand, Chathenlee close, Shell close, Nunne close, Ormlee, Ormesyde.

1956   from *An abstract of the title of Ashdale Land and the Property Co Ltd. to the Acton Farm Estate: Brainshaugh Farm, Ancroft Hill Field, and Connell East Field*:

Part of Heathery Close, Roddings West Wood, Shell's Close, High Field, Arthur's Butt, Arthur's Wood, part of Mill Race Wood, Burn North Wood, Bank, Haugh South Wood and Nursery, Barley Haugh, Burn South Wood, Woody banks, and Waterside House, Riley, Riley Wood, Anncroft Hill, Connel East Field, Far Hag, Middle Hag, Near Hay Arable, Quarry etc., Brainshaugh house, High Chapel Haugh, Low Chapel Haugh.

LOW BUSTON AND WARKWORTH

Low Buston
Low Buston is south-east of Shilbottle, and one original settlement there was a deserted medieval village, with the low-lying earthworks of houses and gardens that flanked the village street still visible. In 1296 there were 11 people paying taxes there. Around 1800 it was turned into a park, with the present house, Low Buston Hall, occupying a site to the north-west. The early field names are of the thirteenth century, including: Bradacre (broad strip of land), the green letch, South Crukes (bends of the stream), the high-rigged acre, Salter's letch, Fletys (streams-fleets), Tyot (?), Allerburn (alders), Salt-rig and Alger-furlang.

Again, the survival of an excellent 1620 map gives a vivid picture of the area. It is part of the Barony of Warkworth; a comparison with the High Buston Tithe map is interesting.

The early seventeenth-century names are here divided in the usual way:

Physical features
Mowdie floores (muddy – 'floores' are areas between ridges in the open-field system), marle pittes, Snapertons (marshy land), Stoneye havers (stony oat-growing lands), wilie dales (willow), sidgedalles (sedge), Broom rigges, Rueleys (rough pasture), Dunston field (stone hill?), Smale latch rigs (small sluggish stream?).

Size and position of fields
The short hauers, The long hauers (oats), Broads, Broad heays (a fenced off part originally), Fore dales, East Waste, The North Croft, Hither Side.

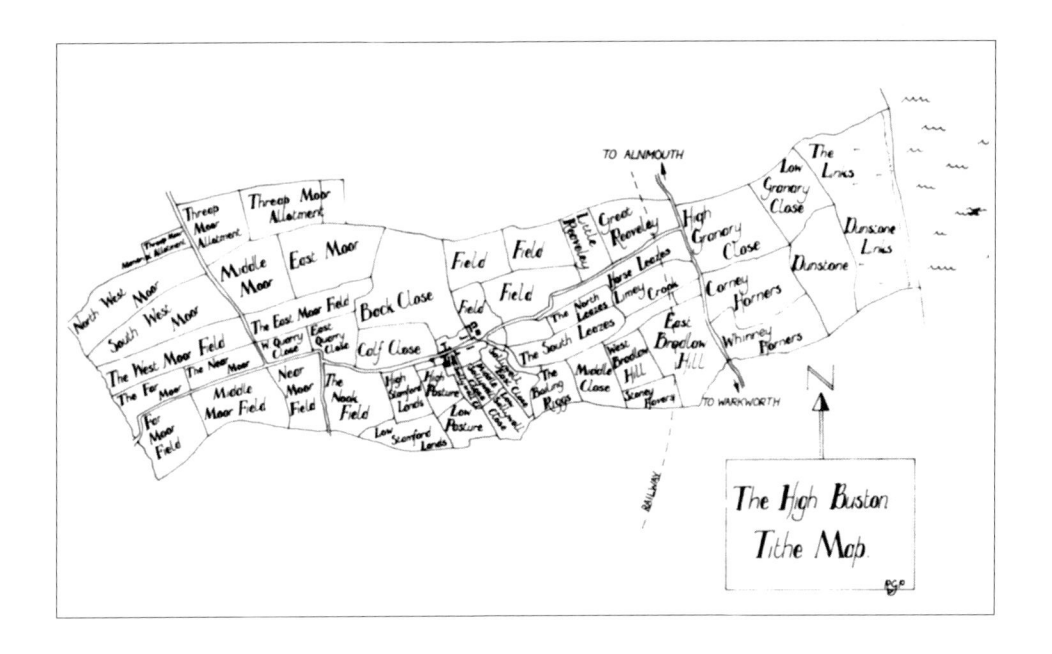

26 High Buston Tithe map (above) and Buston in 1620 (below)

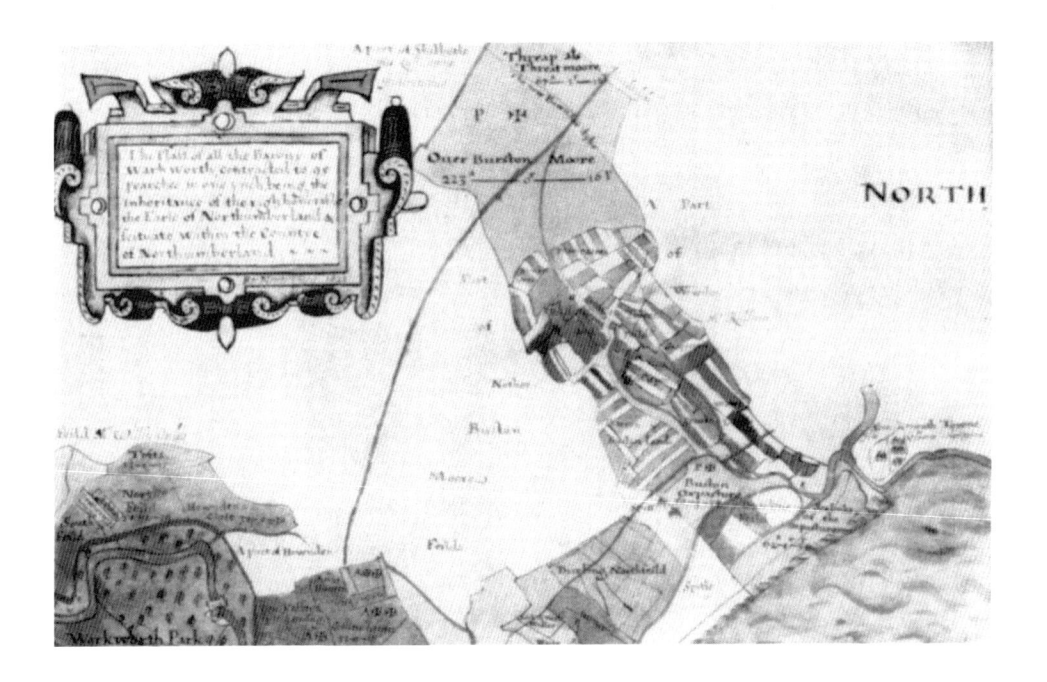

Old field systems and different types of location
The Rood Dalles (a rood is ¼ acre), howmer latch butts, sea butts, Landes, more acer, the flats, haris flatt (hare), West Lawes, Crofts, Sheet havers (oat land parallel strips)

Use of land
Horsley(horse pasture), part of Buston Oxpasture, New rift, howmer meadow, leme lands, and leme crookes (ME *leme*, land that is drained artificially), Stumpert Laws (tree stumps?).

Others
Spitle (belonging to a hospital), Burowlands (belonging to a burgess?), Kilne butts, Threapmoore (disputed ownership).

In 1707 the following names appear: Maddy-rigg (madder was once grown extensively for dyeing), Mill-house rigg, Hounden-mill, (hollow valley), Byar's close, Yard-side riggs, Orchard hill closes, the Bought-riggs (with sheepfolds), Hilly-Law gate, Kideford.

In 1839 survivals include Dunstone, Bradlaw, Stoney Havers, Horse Leazes, Limey Crook, Threap Moor, Allotment and Moor fields.

Warkworth
The Barony of Warkworth is one of the principal holdings of the earls and dukes of Northumberland. In Robert Norton's 1620 map, it is shown lying between Acklington and Low Buston, already examined here. The map shows that the River Coquet, which enters the sea at Amble, changed its course. Warkworth is one of the most attractive

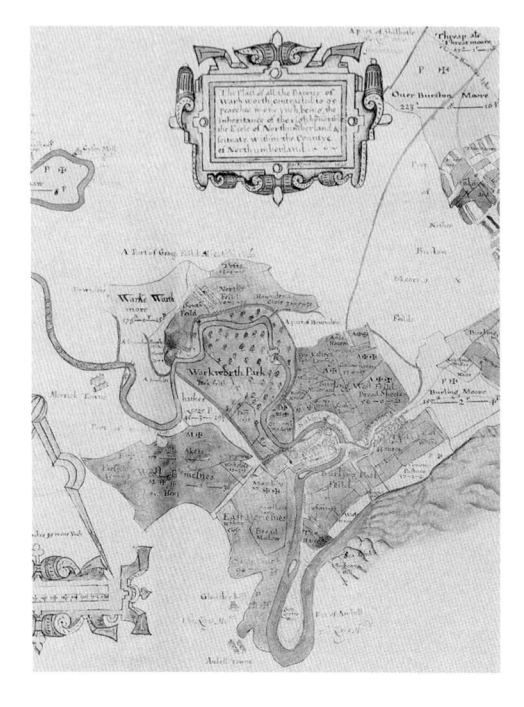

27 Warkworth in the 1620s

villages in Northumberland, contained on a promontory formed by the looped bend in the river, on which the castle stands on a large moated mound. Architecturally, it is one of Britain's finest buildings. However, there is also a fine Norman church at the end of the promontory, a medieval bridge and toll house, and a planned medieval village based on its main road from which burgage plots, or 'scribes' as they are called here, branch out to the east and west.

Field names here have been taken from documents and maps from 1249-1620, which helps to show that some names remain unchanged, whilst some are added and others disappear. They are grouped according to what they tell us about physical features, position and size, agriculture, industry, and social and political conditions.

Physical conditions
Nearness to the sea accounts for Saltgrese in 1471, Salt-grysse (1485), Saltgryse (1498), Salt-gresse (1536), le Salte-grasse (1620), Salterlands (1620), Saut Haugh (1620) and Salt goats (1620). In the last example, 'goat' is a small stream that runs through sand into the sea. Stanecroft (1471) is Stony Crofts in 1620. Waterhewgh or Watershaugh (1576) is on the river, with Hounden-mouth (1567) the mouth of the deep valley. Lez Vyvers (1471, 1485) means the ponds (ME *vivere*).

The following all appear in 1620: Whiny Close, Whinny leaze, East Whins – meaning gorse. There is Hather Leeze and a Thistly Hill. Parkfrith is a wood; there is a Sanding wood, Hesley butts, Hesley Plane and Hesley dales, all named after hazel trees. The haugh, Lowhoop and Cragg bank heads are on the river. Bowell-wele (1485), Bowelhalgh (1495), Bowel-bank (1536), Browel-banck (1607), Bowlbank Leaze (1620) mean that there is a curving boundary (OE *boga*). Whiternes could be from OE *herne*, meaning a white 'nook', or corner of land – the colour of dead grass, possibly.

Position
Welsidmedowe (1471) is the meadow by the spring. Other specific locations are le Pond Close (1607) and Pond Close (1620), and the castle Close (1607). The 1620 map includes South Close, South Field, North Field, Southlands, Kirksyd Bank and Barn Close. In 1567 th'old haven is recorded. Yateside Lonnin (1620) is the track that leads to the gate in the Park fence, and Lain Close is by the lane. Endemyre (1498), Endmyre (1585) and Endemyre (1607) mean the land at the boundary. Warkworth Moore, Brotherwick Bank and Burling Moor (all 1620) take their names from places.

Industry
There was some coal mined locally, which may account for the Orchard pittes (1471) and salt was produced. The other industries named were milling and clothmaking: the milnebatt (1471), Mylnebatte (1485), Milnemedowe (1471), The Milln Batts (1620), Milnsyde butts (1620) and Millerland hauers (1620, now growing oats). ME *teyntour* is a cloth-stretching frame, appearing in Tenter (1620), Tenterhewygh (1567) and Tenter-heugh (1607).

Size of fields

Brademedewe (1471), Brade-meadowe (1485), Broad Meadow (1620) and Broads (1620) are clear enough. Broad Street (1620) means a broad projection and Showlbread (1620) is a narrow piece of land.

Agriculture

Already we have seen the basic units of land cultivation, such as butts and dales. The Orcherd medowe (1471), Orcharde medowe (1567), berne-yerd (1471), le slaughter house (1607), Medow dales (1620), Waste (1620) and Calf Close (1620) identify land usage. Poundeclose (1498) and The pynd-fold (1567) are where beasts were kept. Hungreknoll (1471) was a hill that had poor soil.

Social

Sunderland Park (1249 and 1498) was detached from the Demesne lands, which appear as south maynes (1471), West Maynes Close (in all documents from 1485-1607), Est maynes (1498, 1536), to become West Demesnes in 1620. The Chapel of St Mary Magdalen, which used to stand about 100m north of the present Maudlin Farm, was there in 1200 and mentioned in 1536. Its name frequently changes in spelling: Mawdelyn croft (1471), Mowdeleyn Close (1485), les Mawdelyns closes (1607), Great and Little Mawdlins (1620). St Johns Close (1567) is a reminder that The Knights Hospitallers at Mount St John Baptist in Yorkshire held lands in Warkworth. The church itself gives its name to fords and haughs: Eccleshalghforth (1471), Eglyshalgh (1498), Eccleshalgh-forde (1481, 1536) Eckelhaughe (1607), all making a change from the usual 'kirk' element.

There is Churchwarden lands in 1585, Holylands (1620), and the faint possibility that Harrow Hill and Harrow on the Hill may refer back to a pagan site. The smidy house (1585), the Black Hall (1585), Park Leaze, Common pasture (1620), burrow garths 1567 (land occupied by burgesses) and Hangmanacre speak for themselves. The 1567 survey mentions that newe-towne was originally intended for fisherman and mariners to live in, although by 1567 this was not so. It appears as New-towne in 1607 and Newton in 1620.

Other names

Sperty medowe (1471) is probably from OE *spearca*, covered with brushwood or scrub. There was a burgage or house called Wamboys in 1498, Wamobes in 1585, possibly from OE *hwamm*, a marshy hollow. Headly roods and Headly Hawes (1620) are probably headlands where the plough turned, and the land was good for growing oats ('havers'). Tylbots Close (1607) could either be a building in a small enclosure (OE *teag-botl*) or a building for tile-making. Gallavre havers (1620) might be from OE *galla*, wet, barren land originally, but now growing oats. Houndenz Close (1620) probably means lying in a hollow.

Nearby Amble has these names listed in 1295: The South flat, East Flat, West flat, The Crooks, The Hope, Gonuldes Cross, Dolakelawe, Syket meadow at the Northside hope, The West mede at Blacklawe, The strother.

LESBURY, BILTON AND LONGHOUGHTON

Lesbury

The parish of Lesbury, 4,337 acres, includes the townships of Lesbury, Hawkhill, Bilton, Alnmouth and Wooden. This survey will concentrate on Lesbury township and Bilton.

Lesbury township

This is 1,646 acres and the field names are well covered in documents and maps.

1498   Grysgarth, Rawthornolech, Blackforthland, Redleflat, Middiham flat and Sunderlande.

1567   Swinelee, or Shield-dykes (swine pasture, wall by a shieling), A dyke called Bustone Goate (water outlet to the sea), Doushawe dyke (by a small wood), Rimpeth dyke (the edge of the path), Houghton Besty-forde lande dyke, South flatte dyke, Cottyeards dyke (belonging to the cottars' yards), Chirchakre dyke (church land boundary wall), brocke dyke (brook), Chanley Flatte (later Chanell Flatt on the Aln Estuary), Scanley Flatte, husbandsmans letche, Mere letch (boundary land), Herker Snipes (OE *snaep* is marshy land), Snabs leses (projecting meadow land), Carterdean, Elders Hawghe (elder trees), Broome parke, Calledge Park (crow), Grayst well heade (Grysgarth in 1498, it could be a grass enclosure or paddock at the spring, or more likely a boundary, for 'graye stones' are mentioned as markers), hepstrother hilles (an overgrown marsh, including rosehips), Retche-hewghe (ridge or bank), Hyrde Hill, Sayning banke braye, East and North Seton (coastal settlement), Rose medowe, West noke, Conygarth (rabbit warren), north-west lee rigg, Morysshe buttes (cultivated land by the marsh), abbaye land, Hungere crofte (infertile, needing manure).

The 1567 document shows clearly how many boundary stones were scattered around. A stone stood on the *marche* hill, another stood outside the Snabes leses dyke, and further east were two 'great graye stones'. There were 'thre marche stones sett in the letche beside Rimpeth dyke'. Another word for boundary was 'reane': 'then right doune the north reane of the north west lee rigg...'.

We are shown a detailed picture of problems, for the tenants were poor and the system of strip farming was inefficient: 'every tenant and cottager have in some part of there crofte riggs lyenge amongst ther neighbours', and the reporter wanted them exchanged and the crofts enclosed, but the distribution of fertile ground was too uneven for this.

There was not much stone for building, and natural resources had to be looked after carefully. The reporter mentions ' a good spring of youge allers, yf the same be cheryeshed and hayned and not suffered to be cut down', saying there would be enough in a few years to repair the houses. 'Springe, allers and haynings' appear in field names. After touching on other matters, the writer made the recommendation that the people of Lesbury could improve their lot if they followed a Long Houghton custom: there the tenants were grouped into *ploughe-daylles*, each of which provided a crew for a fishing coble. These tenants had their land 'lyenge rigge and rigge together'. In Lesbury the tenants held their land 'lyeing rigge by rigge and not in flats nor yet in parcels of grounde by yt selfe'. As it was thought that no improvement could be made until the system was changed, a Terrier was made in 1614 to examine the system in detail.

Names of the Westfield parcels

West and East bridge Haugh, Hether Side, Halley Well butts (holy well), Pootes wayst and Pootes lands (poots are unfledged birds and a pullet is a powt, but in dialect 'poot' also means little, insignificant). Broad deales, Cross land butts, Cross land hawuerse, Agnes acres, Durte poote butts, Burn knowle hawyers, Burn knowle roodes (on a hillock by the burn), Earsland roods, Earsland hauers (*ears* = rounded hills, buttocks). In the last name, 'ears' was a common Teutonic name, pronounced 'arse'; thus the Wheatear bird, originally pronounced 'whitearse', suited to the colour of its rump, was changed by a more fastidious people.

North-east field

Long Morrifur lands, Hodden heads letch, Hodden Tippett common meadow, Little Hodden flat, Hodden buttes, Heldon buttes, Heldon hawuers, Tongue buttes (projecting), Hame of Heddons (home), Sweeting roods (pleasant, fertile), Hawuers, West deare sides, East deare sides, Long Weasell Flatte, Griffin buttes, Cross buttes, Hawuers dikes, Hudletch meadowe, Castle close doore, Howle Hungerups (hollow, infertile), Wingyegg lands (windy edge lands), Hanging bauke Hawuers (on a steep slope), Hudletch lands, Cross land flat, Hall knowle roodes, East Hall knowle hawuers, Bancke riggs, Lyme pitt butts, Dungell hoopes, Pinder hill (OE *pundere*, land allocated to the keeper of the parish pound), Pilchesse lands (barked ash), Foure lands, East hawuers, Ruskie hawuers (rushy?), Blinde well lands (hidden spring), Crummy hawuers (Heslop points out that 'crum' means plump or in good condition when used of crops. It can also mean a cow with crumpled horns).

The 1624 map also includes:
Lesbury North Common, Laine butts, Under hill, East Havers, Long Havers, Broom Close, Longfloor haver, Havers Head Law, Short flatt havers, Moores Crooke Roods, Havers Chick-acres, Coathills, Brookes Close, Brook butts, Crookit roodes (on the bends of the stream), Thorney dyke lands, Thorney dyke meadow.

On the same map the East Field includes:
Burn Butts, Westerton Bank, East-bank, West Reans (boundaries), Wha flat (curlew), Wha meadow, Coney Garth (rabbits), Sunderland Havers, Sunderland flatlands, Long Seaheugh, Coatland roods, Alnmouth sides, Wester Seaton, East Seaton, Shellythorns, Crook, Brerland, Keenes acres, Bleasur lands, South floorehavers, Hepies way roods.

Lesbury South Field – seventeenth century map

The names taken from this map have been arranged to show what they describe.
Okletch flatte, Under the Way and Above the Way (oak?), Breamy Hill Havers (*bree* is the breast of the hill or bank; *bremel* is bramble), Spineles (spinney?), Wylie lands (willow), Pan close and leaze (a depression), Water close, Rea close and acres, Cheshill close (gravel), Haugh meadow pasture, Haugh leaze, Littlehaugh leaze, Thorntree lands, Long Sharplaw

PLACE NAMES AND FIELD NAMES OF NORTHUMBERLAND

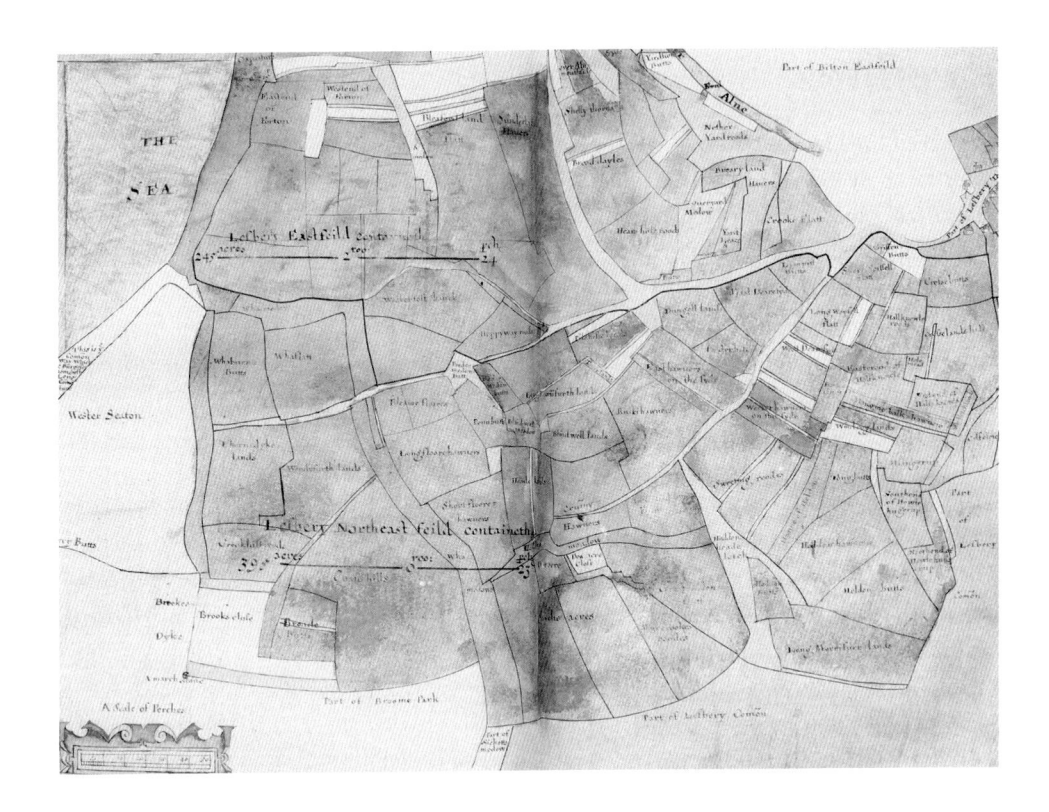

28 Lesbury in 1624: South Field

29 The end of the 'Bounder' document defining boundaries there

Lesbury.

South to the East end of Marden Balke; then west by the South side of the said Balke to the East end of Ware dike betwixt Lesbury and Alnemouth, and soe along the South side of it to a leter which falleth into the water of Alne, And then along the East side of the said water to the west end of the Burrowgh garthes, soe along the west end of the said garthes right South to the Church hill and to the Lowe water marke, then ober the river of Alne fiftie two pearches South, and then turne westward to Buston gate or leter, from thence along the west side of the said river Alne untill wee come to the said Loning w[hi]ch leadeth to Wooden, which is the place where wee begunne.

(steep), Shortsharplaw, Long and Short Mirriches (marshes), Long roodes, Half acres, halfacre butts, hungerfull lases, Okletch Head, Hedland Meadow, Bilton butts, Okletch medow, Acar lands, Todaile medow (fox), Shipsburn (sheep), Lesbury Oxpasture, Delf lands (dug, a quarry), Headlands, Coatland rood, Toddaill butts, Crawlaw (Wester, East, Middle, Greene-), Howlekiln water, Coaly Way, Armorers and Armorer Flatt (surname), Lyle Leaze free (surname), Friskett Roods and Havers (there is a George Fressell and a Francis Freswell).

Botte lands (either butts or a building), Bartholme Butts, Holemeadow and lands, Goodly Crawlaw.

Not all of these names have been explained, as some have already been encountered and the meaning of others is unclear. There is a considerable amount of detail here, making it yet another deep source of local field names.

Bilton

Bilton lies north of Warkworth and Low Buston, on the road from Alnmouth to Shilbottle. Originally it was a village of crofts and gardens fronting on to the main road. The map of Bilton and Lesbury in 1624 is detailed. Bilton has three large fields with subdivisions within them: South, North and East with a variety of names, many going back to the open-field system of strip farming.

Physical features

Hepstrother and Hoppins Greene (OE *hoppet* is a small enclosure), Water-gaites (*gate* is a unit of land, a stint), Flower Riggs, Carricares (*carr* is boggy or rocky; a *currick* is a cairn),

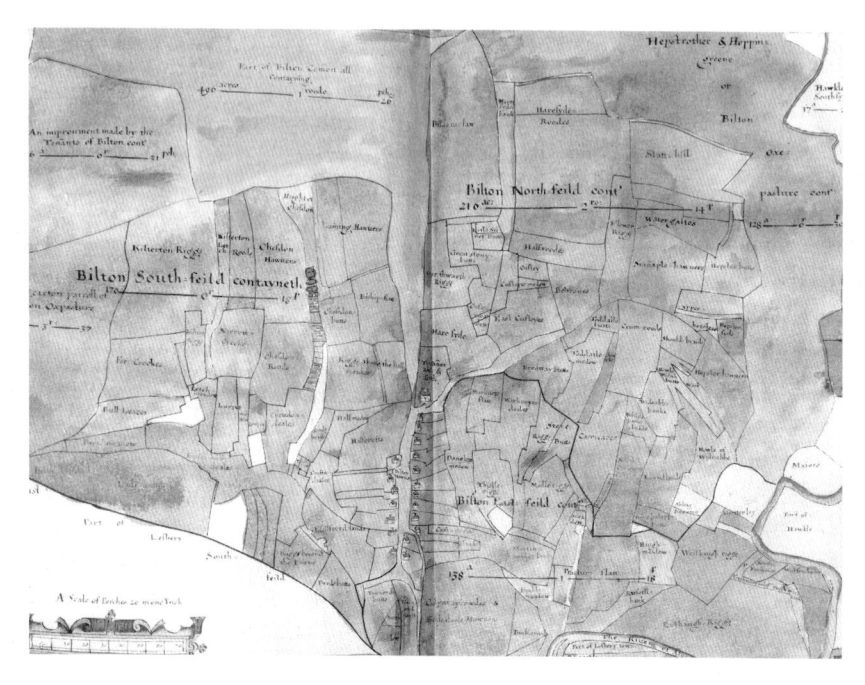

30 Bilton in 1624

Bebdale bankheads, Haughslopp butts (OE *sloh* is mire and *slah* is sloe), Lowndlands (wood?), Upper Hepster (rose hips?), Hepster butts, Thistle Riggs, Russel Bank (rushy?), Haugh Meadow, Broomlands, South Fresh Mire (a freshet is a stream), Freshmire Bank, Chesdon Hawuers, High of Chesdon (gravel), Biship Flatt (OE *busc-hop* is a bushy enclosure in a marsh) and Letch meadow.

Size
Brodsbuts, Broad Meadow, Brodway butts, Mickle daile Howuen, Great stony butts, Little stony butts, Halfroodes, Little Carricares, Narrow crookes, Lang Carton parcell of Bilton Oxpasture.

Position
North Carricares, Westhaugh Riggs, Easthaugh Riggs, Middlerigg, Town end roods, Town end butts, Far Crookes, Riggs beyond the Burne, Riggs aboute the Hall Meadow, Hall medow, Hall crofts.

Agriculture
Overthwart Riggs (a field at right angles to another), Should braids (narrow), Crum roods, Pasture flat, An improvement, Haining Hawuers (enclosed for oats), Croftes-deales, Chesedon-Roods, Broomhill deals, Bull leazes, Chesdon-Roods Havers.

Animals
Hareside Bank and Roodes, Haresyde, Toddaile butts (fox), Toddaile medow.

Industrial
Milne hallands, Coalyway roods.

Others
Martin-bridge land, Girsterley (rented pasture?), Lady Lands, Abbay broome (the abbot of Alnwick held land here in 1498, for example), Warkneys flat and deales (possibly a place where cloth was fulled), Buckle meadow and rood (buckthorn, bent, twisted?) Daneley medow, Slatt Hill.

The *Tithe map of Lesbury township* reveals that the land has been further partitioned, to make way for small fields, many roughly rectangular, with some new elements in the names and some survivals of an older system of farming. North of Lesbury itself are subdivisions of the ox pasture, and an Armourer's Intake. Salkelds Moor could either be a surname or mean willow slope. Bastielands may be from the inner bark of a lime tree, used as a fibre. To the north-east 'Swan' is a surname seen before. There is Hungerup again and many names which show what was kept in the fields, and their positions.

South towards Alnmouth, Haughs appear beside the river, in the bends, thistles, whin, rushes and willows.

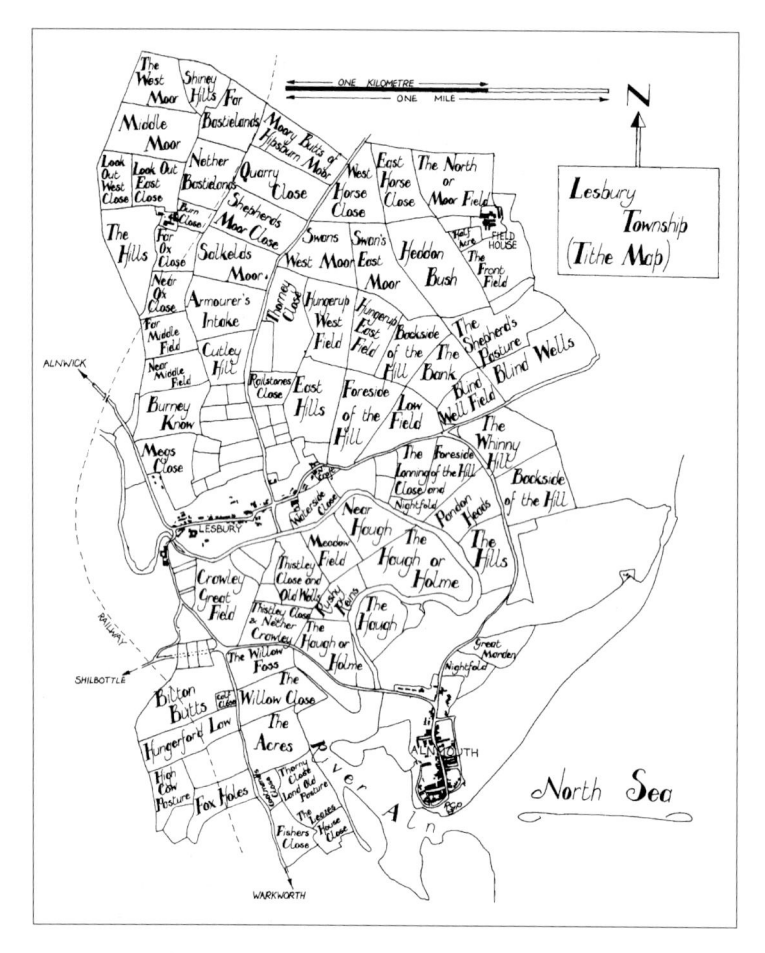

31 Lesbury Tithe map

LONGHOUGHTON

As we move north and further inland, we reach Long Houghton, for which a full survey was made in 1567. It was reported that there were three common fields: East, West and South. Long Houghton was 'a very long town' and the tenants had to travel long distances to their strips of land, so just before 1567 the town was divided in two:

1) West field, north part of the South Field and north part of the East field.
2) South part of the South field and the south part of the East field.

The tenants later woke up to the fact that the southern side was better land, and many complained. The surveyor recommended that a double dike set with quick wood (hawthorn) should separate the two parts, and later maps show that this was done. The report shows how 'sea wracke was useful to the tenants, that the soil was good, pasture likewise and that there were good springs of water in every part'.

The names in this survey were: Lange Hewghe, Chitchacker dick, Connygarth, Sikett dick, Merye butts, morye butt, Brech dick, Bastiford, Est Noyke, Harker Snypes, Sayning Bank Brege, Hirds Hill, Divshey dycke, Rimpeth dyke, Meare letche, Graye-Stone Well-heid,

Snapes Leases, Clatteryng forde, Marche Hill, Kylestone forde, Lucke's crosse, Read forde, the farry rodes, the noyke of Broxfeld Medowe (a dike lately built by John Roddom), the water gait, the houghe, Quarrell hedd, the Quarrells, Kerse Well, the abbay lande, Reverse knowle-yett. These names are contained in a description of the boundaries of the township.

Another interesting feature of the document is the description of how tenants of a plough-dale were responsible for a fishing coble: 'In the antient tyme also the said towne was devidit in plowghdaylles, viz. in every plowghe-daylle 4 tenements, and every tenent had to the same his tenement one cottage appertenyng, and in the same there dwelled one fisher man, so that every plowghe-daylle had one coble goying to the sea.'

A Terrier of 1614 gives more detail, the field names being: Wester dikes, High lawes, South flat, Bastie lands, Law sides, Law heilds (slopes), the flat meadow, South Flase meadow (swampy?), Flase lands, North Flase flat, North Flase meadowe, Middle Flase meadow, North-east Flase meadowe, South Flatt meadowe, Short Aller tofts, Long Aller tofts, Plondon parke, Cleugh meadow, Banke meadow, Wander Knowles, Cleugh Hawuers, Hungery butts, Smoeth (and Smooth) meadowe, Weatifurreans, Kaile yard hawuers, Kaile yard butts, Kaile yard meadowe, Clubshaw flat, Laffer lands, Short hawthornes, South Crum steeles, South Perie acres, Slatt Pitt meadowe, Slatt pitt lands.

What is very obvious from this list is the survival of elements denoting the divisions of the common fields and the variety of descriptive names, some of which are not clear in meaning. In the South Field a collection of local surnames shows how many field names originate – Strother, Bowdon, Shipherd, Carr, Elder, Cleugh, Todde, Rodham, Reade, Hunter.

Individual holdings include Mr George Whitehed's, Sauer meadowe, Steele meadowe, Howle meadowe, Lyllie butts, Easter Seaton Close, Western Seaton Close, Coney Garth, Broom parke arable demeane, Sicketts meadowe demeane, Cotyards demeane. The tenants would have benefited from having their land confined to a more limited area, but it was still divided into many strips, according to the surveyor.

Other names include: Dunshaw pasture, Sea Hewgh pasture, Crawlaw pasture, The Oxe pasture (112 'gaites', one gait being just over an acre), Longhoughton East Moore, Longhoughton North More, Longhoughton West More.

RENNINGTON AND FALLODON

Rennington
Further north-east is the township of Rennington. In the 1622 Survey and Terrier of the manor, the surveyor reports on a survey made in 1618, including these field names:

The South Field: Easter, Middle and West Awards, Unthanke lands (squatters, without leave), North Unthanke, West Dikes, Greeneletch, Old Yard Butts, Hill Flatt, Rennington Orchard.

The West Field: Damheads, Foggie lands, Blindwell Meadowe Butts, Heild hawuers, Long Flatt, Rye Ridds, Blakethin, Linkeyletch meadowe, Howpe Butts, West street lands,

Middle Sheete, Garbutts, Swinter land hawuers, Blindwell lands, Hirst, Short Crofts, West Plowbutts, Short Croft Butts, Renninton leazes.

The North Field: Rockburne Butts, St Mary Knowes, Long Mary Knowes, Harlotte Sheete, Clott Riggs, Fenham Butts, Crofts, Neatherlands, Gewle lands, Weete acres, Brade Arses, Old Yards, Cruckses, Comon Balkes.

Rennington Barelawfield: Barelaw Flatt, Crosse Lands.

Rennington meadowes: Twenty Acre meadowe, Cowde Close, Gowland Croke poole, Orchard Layning.

Others: The Meadowe Dales, Gewle Meadowe, Rock bourne Dayles, Tenement Meadow Deales, The Ox pastures, Sandyeford deane common.

The Tithe map shows how far the township had been divided into smaller units and how

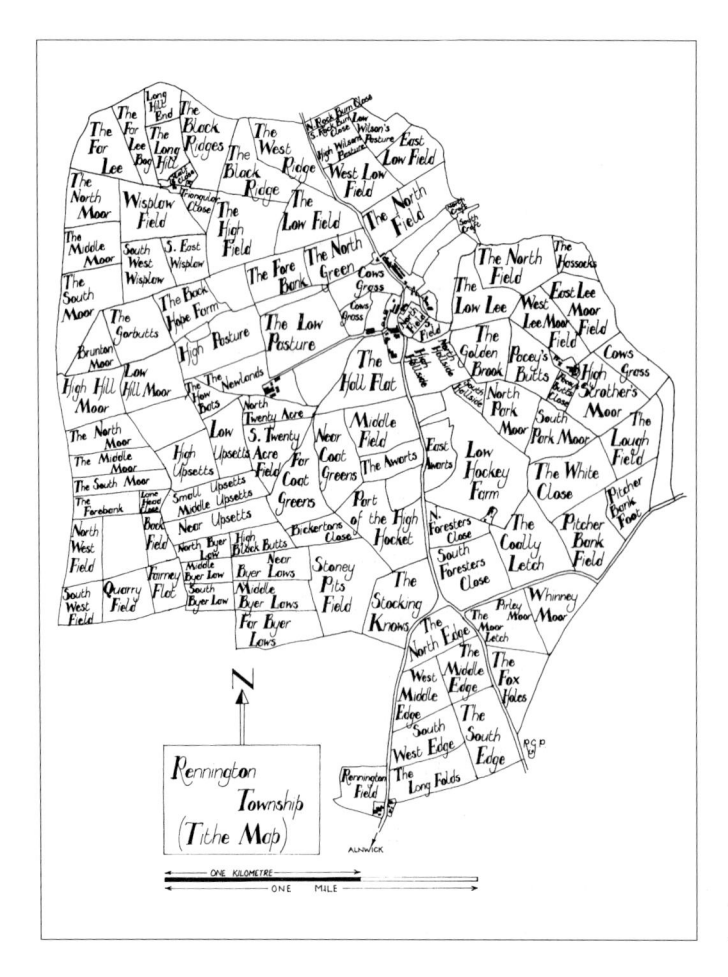

32 Rennington Tithe map

the names have been changed. There are still butts, as in Gorbutts and Pacey's Butts, but the change in naming fields has been fundamental. Keeping to the coastal plain, following it northward, there follows a selection of field maps that draw on many different sources that show how the process of naming fields has developed in the county.

Fallodon
Fallodon here is represented only by the Tithe map. Most of the fields are small enclosures, with reference to gorse, clover, horse, cow and small industries such as coal, bricks and a mill.

ELLINGHAM AND TUGHALL

Ellingham
Still on the coastal plain, but further inland, bordering the A1, the small village of Ellingham was recorded with its lands as part of the Haggerston estate in 1757. To the west, the land becomes open moorland, but elsewhere the fields are mostly regular enclosures.

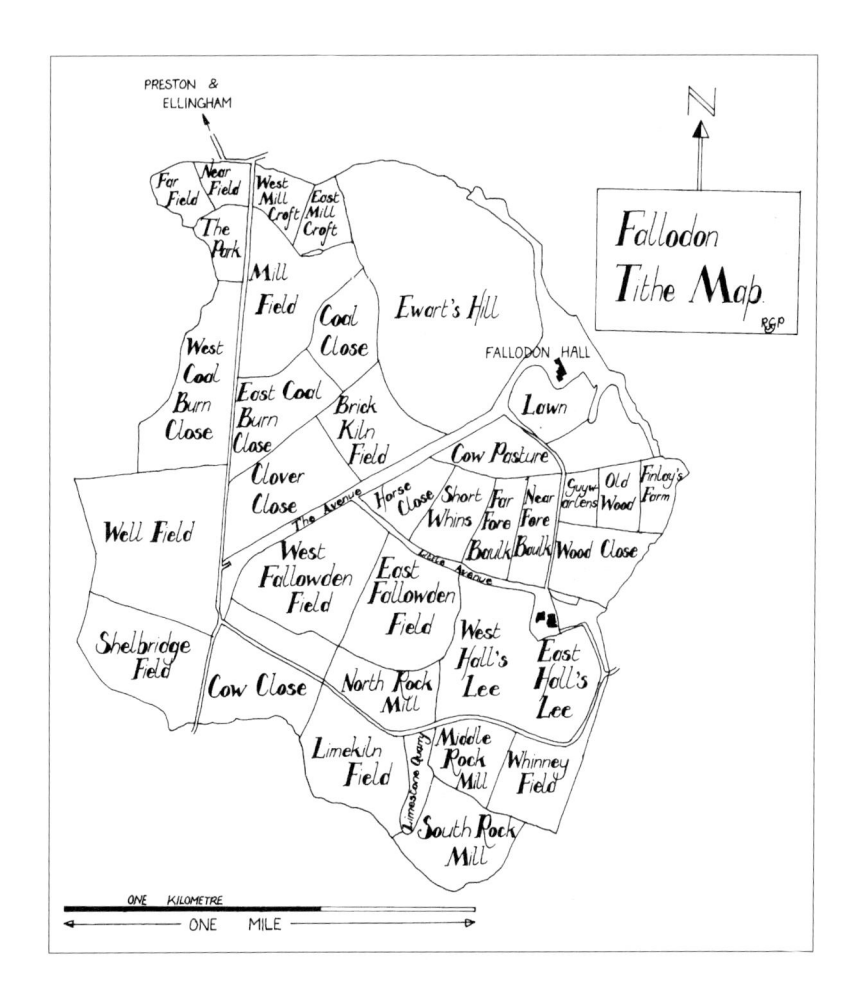

33 Fallodon Tithe map

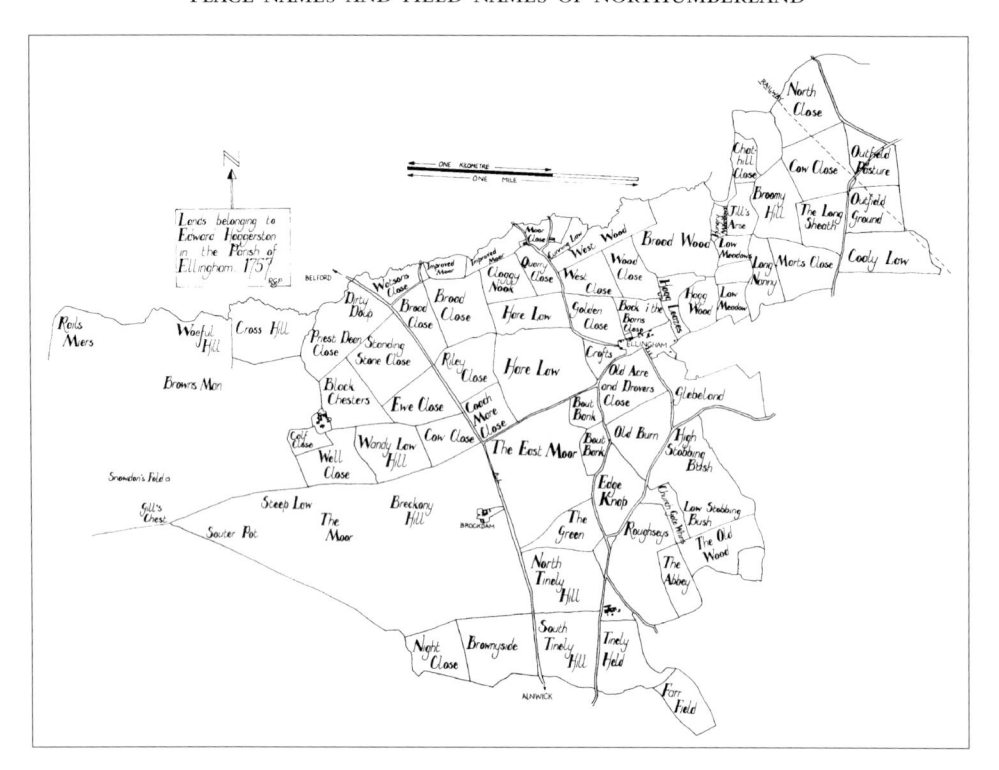

34 Ellingham map, 1757

There are larger fields marked as The East Moor and Hare Law which had probably been brought into cultivation at a late date. The name Tinely is interesting, as it suggests an origin as a field on the Tyne, indicating a small flow of water. Woodland and scrub are evident. Standing Stone Close no longer has a stone there, but this could indicate that one was removed, such as a boundary stone or something more ancient. Black Chesters also may refer to an ancient enclosure. There is a w's Arse and a Dirty Doup, both referring to the shape of buttocks. Riley Close grew rye, and other names suggest an arable function in the old field system: The Long Sheath and the Long Nanny. The Bout is where sheep were enclosed.

Tughall

Two maps have been joined to show changes in names from 1620 to the mid-nineteenth century. Meanings are listed in the glossary.

In *A Bounder of 1567* (Clarkson's Survey) these names appear for Tughall: Sea haughe, Newton scarthe, brasinge brigges, Bruntone longe crofte, Faldar dike, the little moss, the Frears stone, Gibbones dike nooke, the banke nighe the mires, morishe hill, Angry crofte, Wilkins crooke, Turhall haughe, the orcheyard dyke, Small burn, Stories forde, the west burn, Lyherbes leases.

The map shows field names in 1620 and from the Tithe map to illustrate how names change. The western fields are all late enclosures, with precise, straight boundaries in

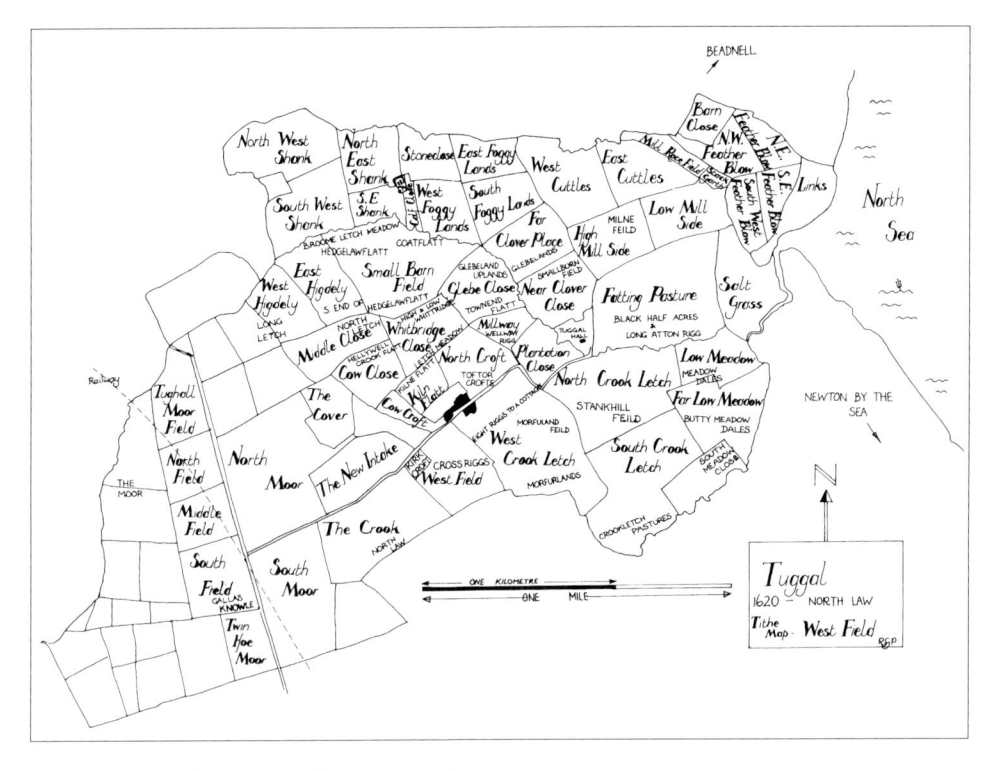

35 Tughall from the seventeenth to nineteenth century

keeping with encroachment on the 'moor', or waste land; the name 'moor' appears, and 'Intake', to show this process at work.

Streams to the south and other watercourses are indicated by Crookletch, indicating how the drainage channels bend. Stankhill is at a pond, and Moorfurlands were marshy. Fatting Pasture contains two earlier fields whose names include acres and rigg from a medieval farming system. Kiln is in constant use and so is Milne Field (High Mill Side). Halleywell, a holy spring, is replaced by Cow Close. High and Low Whittridge become Whitbridge Close. The Glebelands are split up into Glebe Close and Far Clover Place. Hedgelaw Flatt becomes Higdely. To the north are fields named from their proximity to the sea: Featherblow being the light sand. There are Foggy lands, where 'fog' is a coarse, rank grass adjoining fields that lie on the hill shank. Cuttles may be fields drained with deep cuts.

NEWHAM, NEWSTEAD AND ROSEBROUGH

Further north towards Belford is a mid-nineteenth-century mapped group of typical enclosure fields, small with straight boundaries.

A transcript of a document older than 1620 has produced these names for Newstead: Kirkhill ford, Rowledgeford, Parmisford, Proest dean ford, the King's gates, Black

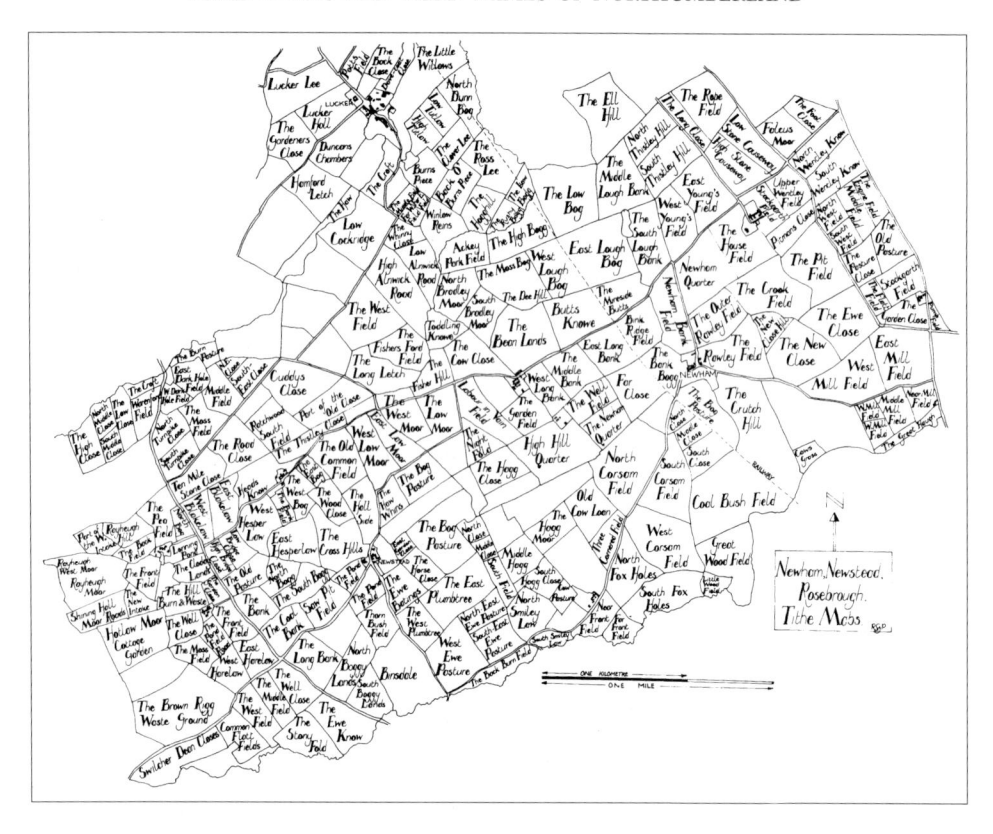

36 Newham, Newstead and Roseborough Tithe map

Chesters, Greenside Latch, Cross Hill, Brown Rigg end, Clattering burn, Black hill, Ray well, Ratchwood, Winlowford, Oaks Scroggs, the Rudford.

Names in 1620 were: Tyl hirn house Pickle, The Oakes, Plumtree Hill, Kirkhill, Kippit law, Hedow Bush, Birkbush, New Ryste, Out Peece, Hoods Knowe, Berry Hill, Blacklaw Leazes, Rosebery Fold, Hesperlaw, Priest Deane, West end of the Demesnes, Demesne Dyke, Switchin Moore deane, Cowstand, Lucker edge, Clatterzan Well, Rail Mires, Rayle Myre, Harelaw, Rownd Know, Warnfoord, In-Moore common, Newstead In, Wretchwood, Easterfield, Paynes Ford, Common Flatt foard, The Maynes, Hinging Well strand, Resshey Pot, Black middings ford, Briggy Hirst.

CHATTON AND LYHAM

Chatton and Lyham lie east of Wooler, on and below the Fell Sandstone scarp that is such a prominent landform as it runs roughly from north to south. Although there is plenty of evidence of prehistoric activity here in rock art, burials and enclosures, the land was marginal, yet suited to a pastoral economy. Arable use intensified only when moorland was enclosed.

PLACE NAMES AND FIELD NAMES OF NORTHUMBERLAND

37 Chatton and Lyham Tithe map

Physical features
Blind Wells, Burn Close, Hetton Close Haugh, Juniper Haugh Waste, West and East Juniper Field, Gilstrother (marsh with a stream), Letches (NE, NW, S, Middle), The Letch Hill, Hollow Burn Close, Spindle Burn Close, North and South Crookey (bends), Birley Hill End, Birley Bush, Thorlow Wood (High, Low), Fairney Law (ferny hill), Thorn Close, Broomie Field and Law, North and South Tanzy Crooks (a herb), Stob Ridge Close, Midge Haugh.

Shape and size of field
Ten Acre Close, Ten Rigs, Quarter Field, Triangular Close, Little Close, North Long Close.

Position
Stone Bridge Close, Face of Park Hill Road, Far-Near-Back Close, Field (West (2), South, East), North Field Pasture, Nether Close, South Close, Head Close, North Close, Well Close, North Well Close, Lyhan Lee, Lyham North Moor, Lyham South Moor, Hetton Close, Corner Haugh.

Agriculture
Rye Hill (N, S, Low), Crop Close, Garden Close, Clover Close, Stackyard Close, Low Close, Calf Close, Ox Close, East Ewe Pasture, West New Pasture, Inclosure (several), New Intake, New Ground, New Close, New Ground, Rough Close, Lee Close, Tetherley Balks.

Animals and birds
Dove Cote Lee Closes, Henlaw (NW, SE, SW), Cat Law Meadow.

Industrial
Dam Close, Mill Close, Old Mill Piece, Millers Intake, Mill Burn Close, The Mill Burn Field, The Peat Holes.

Personal names, social
Alice's Close, Laidler's Haugh and Half Waste, Luke's Haugh, North and South Priests Meadow, Priests Haugh, North and South Slaters Lee, Jobs Fields.

Others
Brandy Well Close, Marlins Bush, Harnish Hill, (mill?) Pond, The Glander Close, Hingston Lee, The Lawns, Lawns House Field, Curls Walls, West Middle Clarilaws, Wandon Lowfield and Know, Blindy Bush, Piper bank, Reddishall, Middle and West Reedis (reeds?).

There is an extraordinary number of narrow, unnamed garden-like strips around the village of Chatton, many of which border the River Till.

38 West over the New Intake and Park Hill, over Chatton to the Cheviots

Place names

Chatton	Chetton	1178	OE *Ceatta-tun*, Ceatta's farm
Lyham	Leum	1242	OE *leah-hamm*, grove meadow
Hetton	Hetton	1163	OE *heath-tun*, dwelling on a heath

LUCKER AND DETCHANT

Lucker lies to the north, with these names given in 1620: Gawland Quarter, Broomhill parcel of the Oxpasture, Mooly fludds, Benfield dike rigs, Blind Wells, Bradford dyke boggs, Rippitlaw Chesters, Grind Lands, Black Briggs, Newford peece, Flatt of lands, Bentlands Close, Bean lands, Reane lands, Izlands, Nether Cruicksses, Woodwaies, Gilders, Bank Quarter, Crankesses, Bank riggs, Crofts, Piperside butts, Hill flat, Long lands and Short lands, Tuttlaw Close, Broome Close, Broom piece, East coat close, Northbroomcoat Knowes, Dove Cote Close, Green Orchard, South cote close, Gaile Close, Southlough close, Northlough close, Town Dales, Westfield, Crofts, Shipp coats, Knotforth foord, Short and Long Knotforth, New rift, Woumlyhill, Long Acres, Upper long acres, Cockeridge, Aple garth, Blackburn Havers, Well hawgh, Peter lands, Butts, Kitwell Lough.

The Detchant Estate map of 1858 covers land close to the North Sea, flanked by the Buckton Burn to the north and moorland to the south-west, the latter with coal and lime deposits that have been exploited, as South Kiln Field, Colliery Engine, Pit House and Coal Houses Field indicate.

Kettle Burn Farm has its fields numbered on the same map, but not named. The regular boundaries of fields are determined by enclosure, but those boundaries that follow streams and rock outcrops are irregular. There is wet land, with names such as Weetside and Bog Field. There are ponds: Swinhoe Pond Field had pigs and so did Swinhoe ridge. Is Fawcet Field connected with water or a personal name?

There are 'Haughs' beside the stream, The Brockses are named after brooks and Blakely Willows has a pond at its centre. There are Mill Fields. Stackyards, Stackyard Field, Fattening Pasture, Broad Meadow and Old Cow pasture speak of animal husbandry. Henshope Law may be a hill where there were hens or game birds. Grey Mare Field may be one with a boundary marker. Another field had a Broken Cross. Virgin Hill may have been fallow, but Upper, Middle and Lower Slashes were in cultivation.

Fields with owner's associations are Dick's Old Walls, Squires Close, Clarks Close and Forsters Close ('forester's'). Hope's Ways presumably refers to a valley, but a personal name is possible.

SCREMERSTON, SPINDLESTONE, OUTCHESTER AND BAMBURGH

So far, information has been based mainly on the Duke of Northumberland's estates. There is an important source of field names known as the Greenwich Hospital maps, which deal with the confiscated estates of the Jacobite the Earl of Derwentwater, executed for his part in

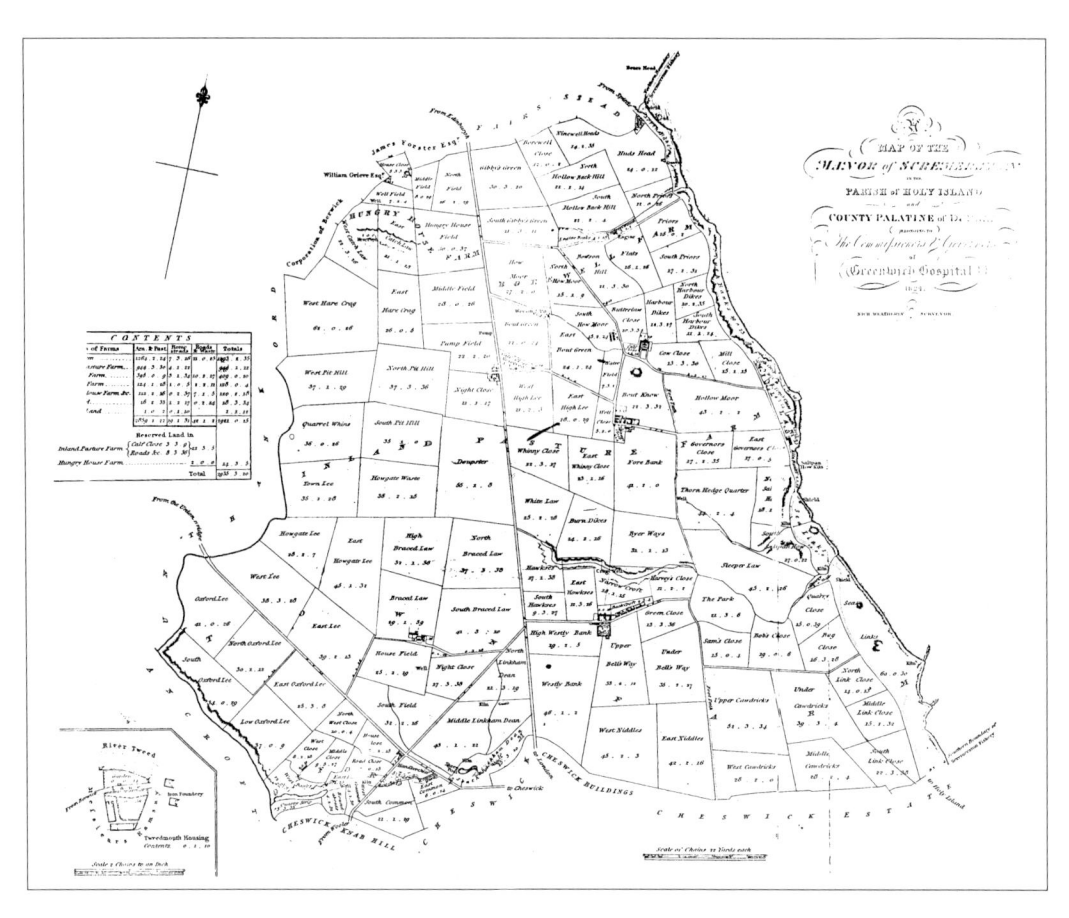

39 Scremerston in 1824

the Rebellion. The lands were meticulously surveyed and show every acre, rood and perch as well as the field names. Many of these lie in the Tynedale area; the Earl's family base was at Dilston Castle. There are some outliers, and two of these are chosen: Scremerston, in the far north-east of the county, and Spindlestone and Outchester in the Parish of Bamburgh.

Scremerston

The Great North Road divided the area from east to west. The fields are typical enclosures with straight boundaries. Hungry House Farm lies in the north-west corner, possibly meaning grassland or not very fertile ground. One cluster of fields refers to small industries: pits and millstone-extraction, others to waste, animal enclosures, an ox ford, and there is the intriguing 'Doupster', generally meaning buttock-shaped (and 'doup').

East of the road there are quarries and links close to the sea and the interesting field called 'Niddles', which may be dales, strips of land going to a point, but this is speculative. The possible meanings of others are in the alphabetical list.

Spindlestone and Outchester

The area includes outcrops of whinstone which provided quarrying and sites for small defensive prehistoric enclosures. Spindlestone itself is the site of a legend involving

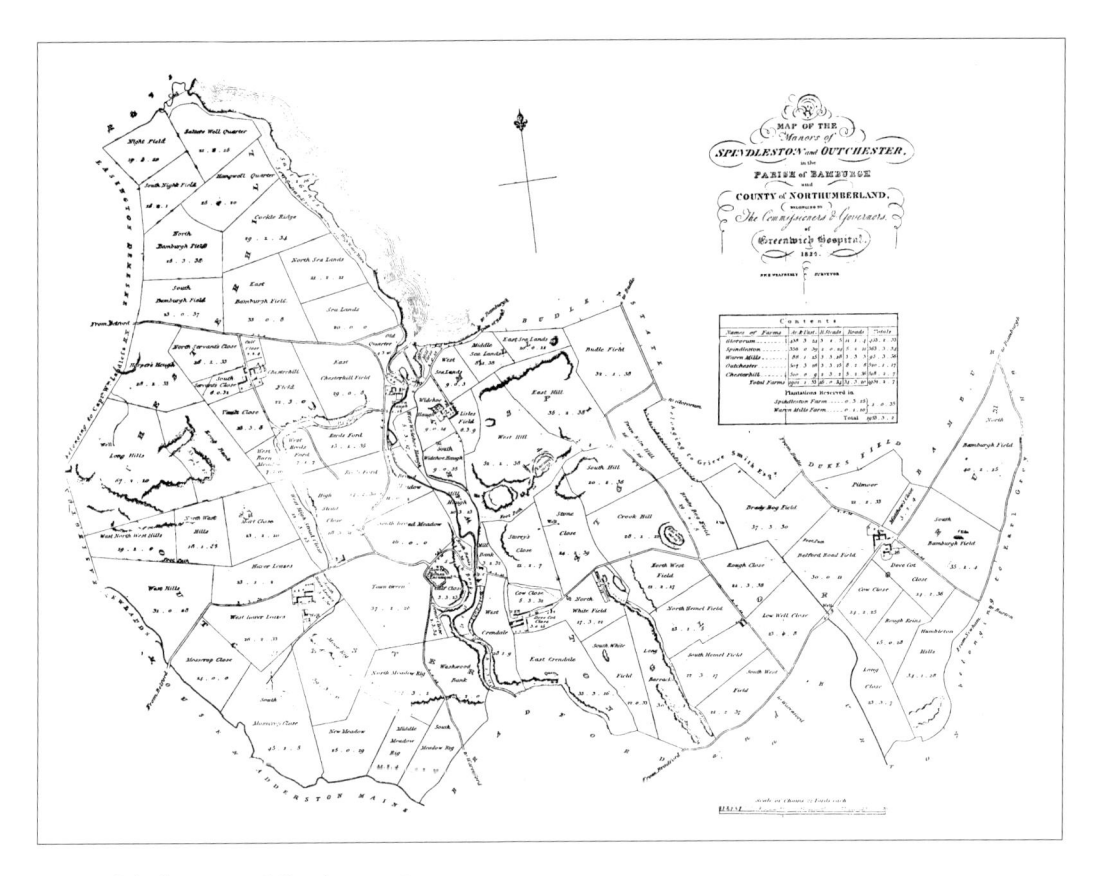

40 Spindlestone and Outchester, 1824

wicked witches, a noble young prince and his sister who was turned into a dragon for being regarded as more beautiful than her witch-stepmother.

The whin dykes give rise to many names containing 'hill', including Humbleton which means bare-headed. Sea Lands border the North Sea. 'Haver' and 'Leazes' (oats and meadows) have survived.

Some of the earlier field names of Outchester in 1577 are: Tenter Bank, Deans Mill, Walk Mill (all connected with cloth making), Pavyne, Moscroppe, the new meadow, Cross Flatt, Read Ford, Stonylands, Grange path, Connrigs, The Cross Loan, Cald close, Long Elmes, Stuffleu Hill-Stifley Hill, The Great Hugh or Cragg called the West Heugh, Paddowell, Harpers heughe, Ross Dyke, Sclate Burn and Slaitburne, The mill meadow, The mill Willway, the Buts, Burn Crook.

Bamburgh

Bamburgh Demesne Lands are listed in 1334 as: Netheredlange, Querred langes, Rankstaneflat, Horslawes, Katakre, Estcrosflat, Westcrosflat, Sewleys, Baserflat, Swanlawflat, Northfeld, Quarelflat, Blyndwellflat, Shelrygge, Sokflat, Grenewellflat.

Land belonging to Bamburgh Friars in the seventeenth century consists of: The pasture part, The porter's or potter's field, West green Close, The Black Hill, Ragflet, The east green Close, The Hornes, The helpits, The helpit Close, Threap Ground.

To balance the coverage of names, three other areas in the northern part of the county are now chosen as representatives: Pawston, Alnham in the Cheviot Hills and a group in the north-west from Elsdon to Rochester. They cover different kinds of land in height and soil types and in some ways contrast with the fertile coastal plain. All are based on information from Tithe maps.

PAWSTON AND ALNHAM

Pawston

Pawston lies close to the western border with Scotland, a small village in the Cheviot hills. As a sample, names on Pawston Farm in 1899 are given: West Carr Close, Great Sloshes, Little Sloshes, Smithy Field, Mill Haugh, Carr Haugh, Ash Trees, Well Field, Croft, Ponitons Close, Hobb's Croft, Pillmoor Haugh, Pillmoor (NW, SW, SE and NE), Cauld Back, West Baulks, Wheat lands, Poned lands, Pawston Hill, Bright Knowes, Shepherds Field, Ewe Field, Notley Know, Botany Bay, North Hill, South hill, Back Hill, Shildene, East Hill Slant (?), Road Field, Kilham Bank, Riverside Pasture.

Alnham

The area covered has 9,535 acres and 10 perches. The field names that follow come from the 1843 survey, including the farms of Northfield Head, Penny Laws, part of West Unthank Close, The Castle Farm, Alnham Home Farm and Hestleton Ridge Farm. Alnham has a long history of settlement from prehistoric times onwards, with

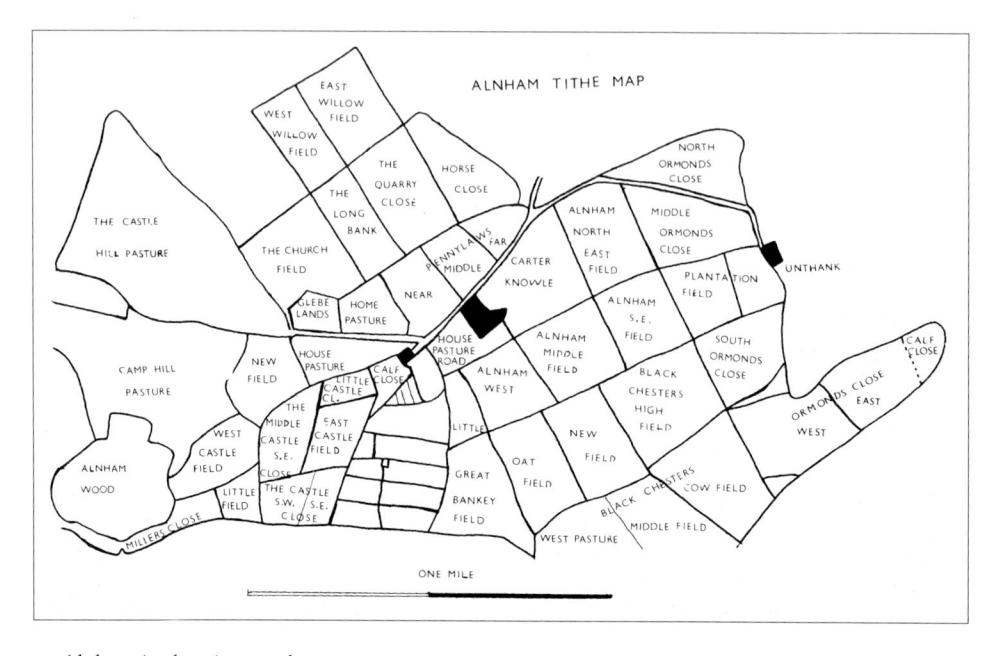

41 Alnham in the nineteenth century

burial cairns and pre-Roman settlements in the hills, notably at High Knowes, and a dominant long-used fortification called a castle. However, there is also a restored medieval fortified tower house next to the old church; this faces a field in which there is a prominent grassed castle mound mentioned in 1405, and burnt down by the Scots in 1532. The word 'castle' thus has extensive use. The church shows signs of being the centre of a declining village, for the north aisle was taken down and the arches filled in to make it smaller.

The field names are: West and East Willow Field, Little Castle Field, The Long Bank, Great Bankey (on a slope), Hestleton ridge moor (Hazel), Black Chesters High Field, West and East Castle Field, Black Chesters Middle Field, The Castle SW Close, Penny Laws Near, Middle and Far Fields, Alnham Moor, The Middle Castle SE Close, Alnham Moor East Field, South East Field, Middle Field, West Field, Home Pasture, House Pasture, The castle Hill pasture, The Lea Field, Pasture, New Field (2), Camp Hill Pasture, Plantation Field, Black Chesters cow field, Black Chesters West Pasture, Horse Close, The Quarry Close, Alnham Glebe lands (church), The Millers Close, Ormonds Close (N, S, E, W, mid-), Unthank West Wood (disputed ownership or having squatters), Cobden Pastures (top of a hill), Carter Knowle (The Salters Road runs through here), Penny Laws Near Field (rent? small pieces of land? coin hoard?). The tenants' names show the continuity of local surnames: Storey, Tully, Green, Fettis, Hindmarsh, Crawford, Awburn, Atkinson, Chrisp.

ELSDON, MONKRIDGE, WOODSIDE, TROUGHEND, OTTERBURN AND ROCHESTER

Two abbreviations used here are: A for 'Ancient Land' and C for 'Common Land'. All fields are both Ancient and Common, unless A or C appear separately. All are from the nineteenth century. Meanings are given in the general index.

42 Signpost: Elsdon area

Elsdon Ward

Elsdon has a very large village green, reminding us that this was a focal point of drove roads along which cattle were driven from Scotland to fatting pastures and markets in England. Dunshold and Low Carrick, East Toddles (fox hills), North, South and Middle Riding (clearing), High Moat (of the motte and bailey castle), North Bower Shield, A and C land at Elsdon, Crown Inn, Close and Haugh, Landshott, N. Whitlees, Cat Pool farm, Batt (A), Town Head, Sandybrae, Bitchpool, Four Riggs, Burn Crooks, Dunsdole, St Mary Well, Hill Head (C), Hudspeth, Landshott, South Bower Shield, Moat Hills (A), Flatt Fell, Land in Elsdon Bog, Whiskers Shields, Harthouse, Hudspeth (path), East Nook and Colsters, John Croft and Spartishaw, Elsdon Gate Cheek, Bainshaw Bogs (C), Redshaw, West Tod Holes, Bird in Bush Inn, Glebe Lands, Whitlees and Lea Houses, House and garden at Elsdon and Star Myres, Dixons Hole, Elsdon Mill Lands (C), Scotch Arms Inlands and Lonning house, Burnstones and Low Moat, Landshott End (A), Town Foot (A), Pearson's House land, Hill Head (C).

This impressive list of names contains many that hark back to the rig and furrow systems of arable farming (batt, shott, riggs, flatt), and in the area there are still traces in low sunlight of such systems to the south and south-west of the village.

Monkridge Ward

Year haugh, Halls hill, Raylees (4), Dyke nook, Raven Cleugh and Burn head, Rayless (C), Dykenook, Dykenook Green (C), Monkridge, Knightside (A), Hole, Haining (A), Ottercops.

Woodside Ward

Craig, Hedshope (5), Raw (A), Grasslees, Grasslees Mill or Hill?, Wainford Rigg, Highshaw and Ironhouse, Laingshill, High Carrick, Dunns Loning Burn and Pauchfords, Deer Park, Herdlaw, Brockley Park Rimpide and Plantations (A).

Troughend Ward

Woolaw, Bog (A), Woodhill, Birdhope, Garrettshields, Blakehope, Greenchesters (A), Townhead Haugh (A), Rattenrow (A), Troughend and tofts, Old Town, Evestones and Netherhouses (A), Smartside (A), Ash-trees (A), Kelly-burn (A), Cleughbrae (A), Brownrigg Common (C), Netherhouses (A), Evestones (C), Earls meadow (A), Rooking, Troughend and Garretshields Allotments (C), Chattelhopes Babswood (C), Old Town, Meadow haugh, Raw (C), Blackblakehope (C), Blakehopesburnhaugh (C), Whitelee (C).

Otterburn Ward

Hatherwick (A), Wainfordrig (C), Colwellhill, Greenchesters (A), Townhead, West Cottage farm, Otterburn Farm (A), Shittleheugh, Easthopehead (A), Daveyshield Hill (A), Hopefoot (A), Daveyshield (C), Disputed land (C), Otterburns Girsonfield, Mill Farm (C), Closehead, Monkridgehall, Over-acres, Heatherwick, Pottsdurtrees, Toft Ridge (C), Soppit, Carrick Bushes (A), Garden at Monkridgehall (C), Westhopehead (A), Garden at Fairneycleugh (C), Fairneycleugh, Branshaw, Disputed land (C).

PLACE NAMES AND FIELD NAMES OF NORTHUMBERLAND

43 Newlands and Whittonstall. *Greenwich Hospital*

Rochester
Birdhopecraig, Hillock, Silloans Common Land, Dykehead (A), Low Rochester (A), Lumsdon Common Land, Featherwood Common Land, Elishaw Mill(A), Bellshield, Sills, Rochester Peel (A), Low Rochester, (A), South Chester (A), Byrness Glebe (C), Cottenshope, Tofthouse, Catcleugh and Spithope Common Land, Durtrees and Murridge House and Yatesfield Wanlaces, Disputed land common land, Byrness Common Land, Cottonshope Burnfoot (C), Horsley, Stobbs, Bagraw, Stewartshields, East Dudlees, Elishaw (A), Elishaw (C), Birkhill (A), W. Dudlees, Dudlees (C), Storage, Low Rochester and Petty Knows (A), Rochester Common.

SOUTH NORTHUMBERLAND, WITH NEWLANDS AND WHITTONSTALL, DILSTON, THROCKLEY AND COASTLEY

The final section of this survey is covered mainly by the Greenwich Hospital maps of the confiscated estates of the Earl of Derwentwater. These are standardised and clearly marked in acres, roods and perches, with each field named. Many need little 'interpretation', as their meanings appear to be clear, and as the names are included in the alphabetical list, the maps themselves are left to do the work.

As an example, I begin with *The Newlands and Whittonstall Estates (1824)*, which lie to the south of the River Tyne, centred on the Roman road, Dere Street, that ran to Corbridge and beyond (see p127).

The field patterns are regular, typical of late enclosure. The physical features depicted include Crooks (bends), Letches, Sykes, Haughs, Bog Fields and Wet Field, Redway Wood (reeds), Little dean, Gillside, The Mires, Blackburn and Water Way. The vegetation is alders (Allery), Thistly, Broomly, Hazley, Whinney, with The Hagg, Scroggs and Birchy Pasture.

Hills, fells and banks are named, and there is the usual number of names that show position in relationship to each other and to other features such as Path Head, Above Town, Over Hill and under Field.

Agricultural practice is well represented by pastures, rigs, meadow, stubble, haystacks, barns, cows, calves, oxen and geese. There are oats, peas, wheat, rye and rape. There is a Fallow Side, a Big Stubble Syke and a reminder of an earlier system in Bank End Butts, North Dales and The Riggs.

Industries are represented by Colliers Field, Kiln Pit, Quarry Banks, Mill Hills and Salters Field House. There are Glebe Field, Shotley Church Field, Holy Well Field, Thompson, Clark, Blackey, Hamilton, Smith, Gibson, Brown, Steven, Abraham and Wilkinson. Watch Hill describes its height and position, Grey Mare Hill is probably a boundary marker, Currick (Close) is a pile of stones. Tougham, Howney, Wrey, Hepper Mill, Bullionside and Rossle Side have their counterparts, to be found in the index.

Above and next page: 44 The South Tyne in the 1820s

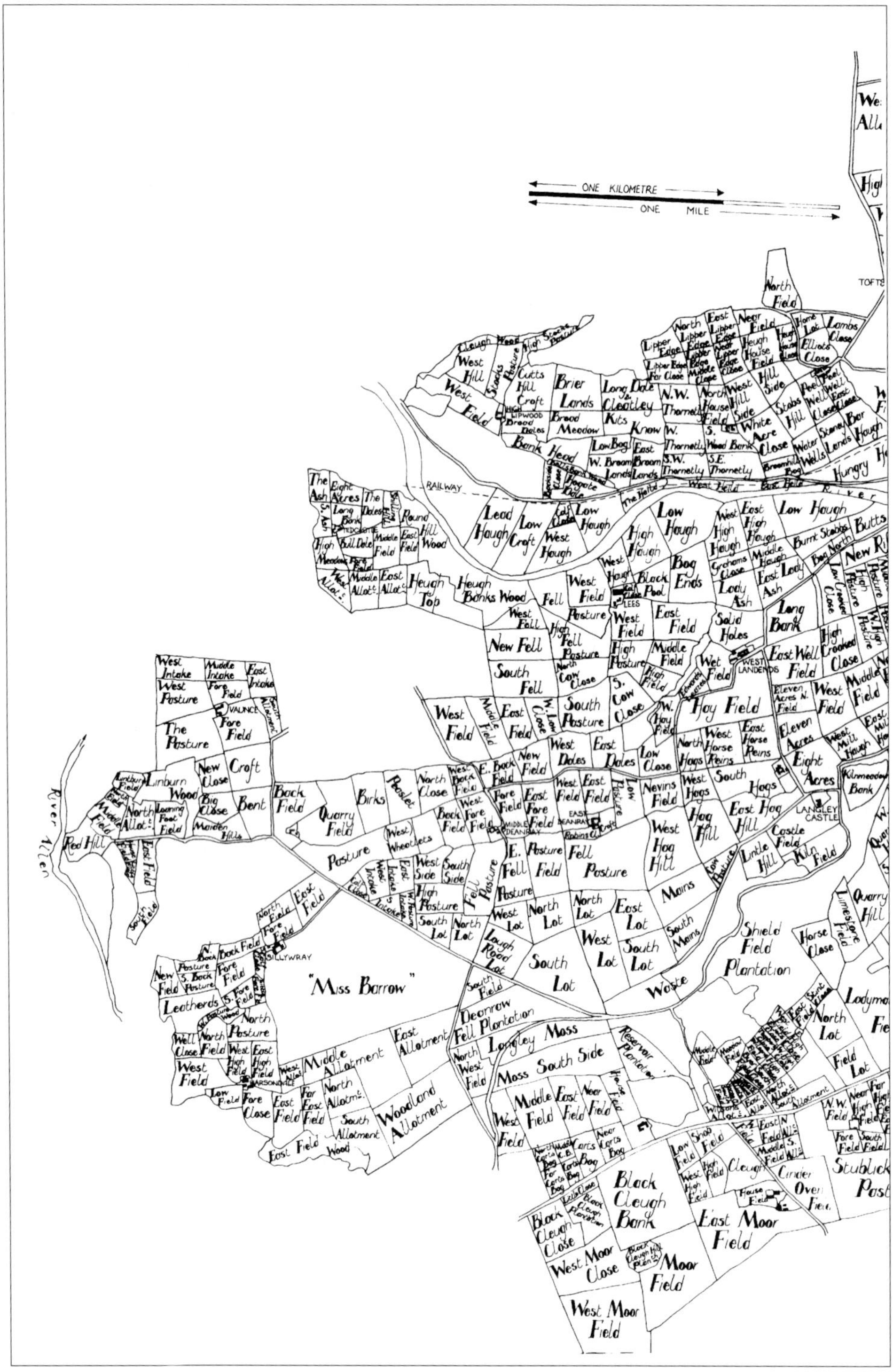
ONE KILOMETRE
ONE MILE
RAILWAY
River Allen
"Miss Barrow"
LANGLEY CASTLE
SILLYWRAY
LEES
WEST LANDENDS
TOFTS
Lead Haugh
Low Croft
Low Haugh
High Haugh
Bog Ends
Black Pool
Heugh Top
Heugh Banks Wood
New Fell
South Fell
West Field
Fell Posture
South Posture
Low Close
Hay Field
Eight Acres
Eleven Acres
Long Bank
Horse Close
Quarry Field
Birks
Back Field
Bent
New Close
Big Close
Red Hill
The Posture
West Intake
Middle Intake
East Intake
West Posture
Shield Field Plantation
South Lot
West Lot
North Lot
Mains
Waste
Langley Moss
Moss South Side
Deanraw Fell Plantation
Middle Allotment
East Allotment
Woodland Allotment
South Allotment
Black Cleugh Bank
Black Cleugh Close
East Moor Field
Moor Field
West Moor Close
West Moor Field
Brier Lands
Broad Meadow
Long Dale
Cleatley
Kits Know
West Hill
Lintle Hill
Kiln Field
Castle Field
Hungry Hill
Butts
Stublick Posture
Leatherds
Pasture
Burnt Stobbs
Low Hough
North Field
Lambs Close
Elliots Close
Cinder Oven Field

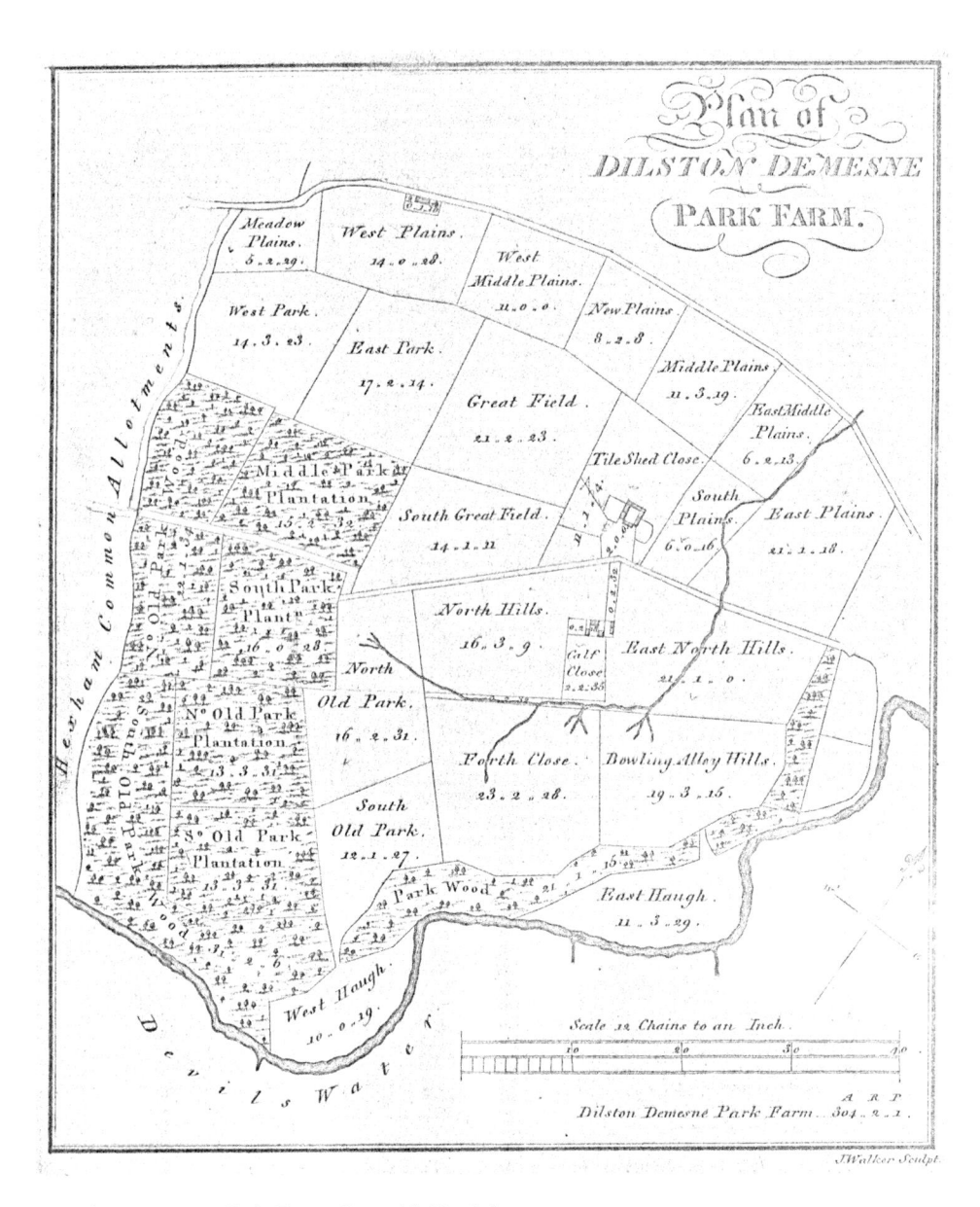

45 Dilston Demesne Park Farm. *Greenwich Hospital*

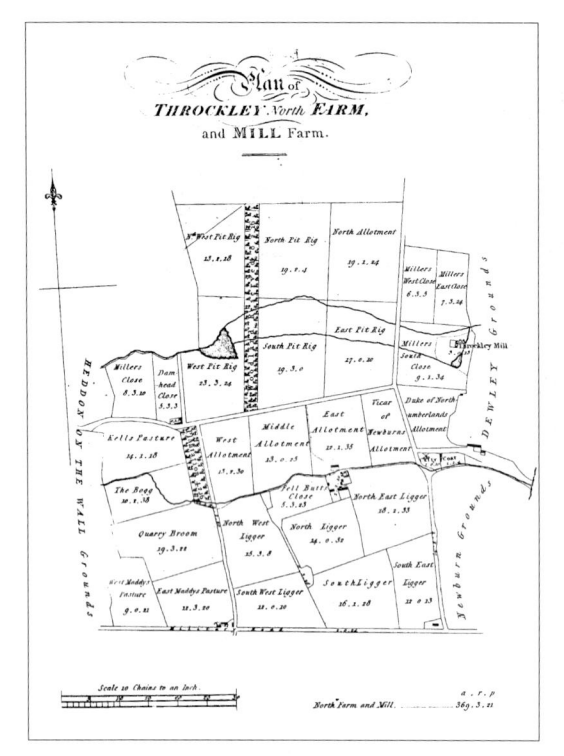

46 Throckley North Farm. *Greenwich Hospital*

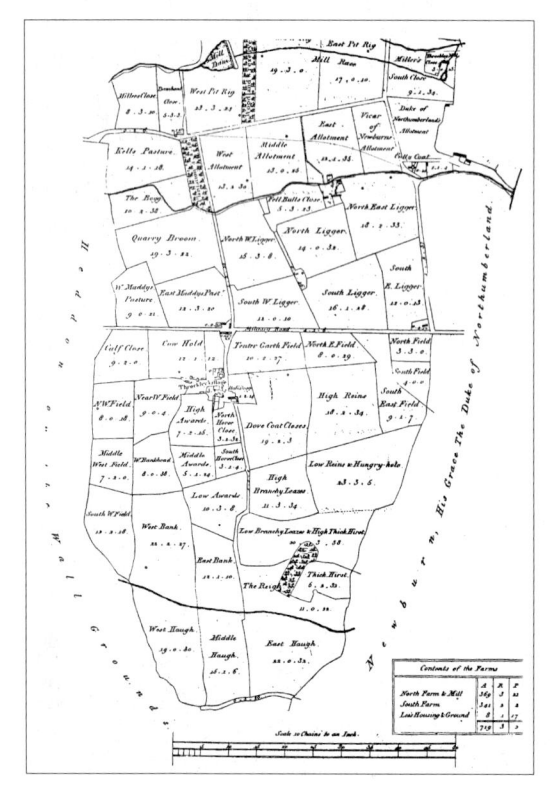

47 Throckley Mill Farm. *Greenwich Hospital*

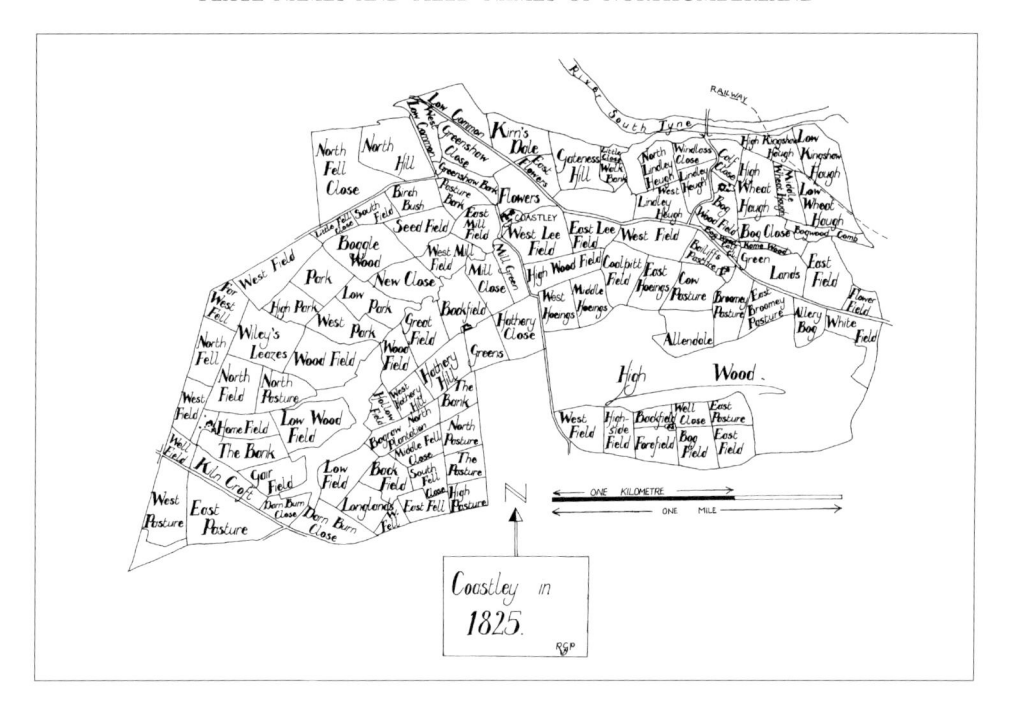

Above: 48 Coastley in 1825

Right: 49 South Middleton deserted village: a fossilised ancient farming system This shows very clearly the ancient 'strip' system of agriculture within the large fields, here eventually abandoned for pasture. Streets, house, gardens and plots are still visible in the blown snow. *Museum of antiquities, Newcastle*

THE SOUTH TYNE AREA

The map produced on p129-130 is a compilation of several individual maps of 1825. The map of Dilston itself, power-base of the estates, is included on p131. The total estates amounted to 2,272 acres. Throckley is represented by two maps, reproduced on p132, now administratively in Tyne and Wear.

Using information from the same survey, the map of Coastley in 1825 (p133) has been redrawn by Richard Parkin to the style of the others in the early part of this book. The acreage has been omitted. The photograph of South Middleton village on p133 very clearly shows the ancient 'strip' system of agriculture within the large fields, here eventually abandoned for pasture. Streets, house, gardens and plots are still visible in the blown snow.

ALPHABETICAL LIST OF FIELD NAMES

(a) Some of the early dates are included with the examples to show the range of usage.
(b) The list excludes most definite personal names.
(c) The denominatives are not always given: they are to be found in Part 1.
(d) Alternative derivations are given to assist the reader with his or her own discoveries. Some possible explanations are, of necessity, tentative.
(e) Instances of H or (H) refer to the work of Oliver Heslop (see bibliography).

(the) abbay lande (1585): land belonging to the abbey
Acar lands (1624)
(The) Acre Close: OE *aecer* = plot, acre
Adlew Close: personal name?
Aikeley Park (1620): the clearing in the oak wood
Aldchestre (1479): OE *ald-ceaster* = old fortification
Alder Bank: OE *alor* = alder
Aller Tofts (1614)
Allery Bog
Alicot
Alnmouth sides (1624): place; OE *side* = land along the river
(Chester House) Allotment
Angry Crofte (1620): OE *anger* = pasture, grass land
Anger Croft (1567)
Annaysacre (1250): personal name

Antecroke meade (1620): meadowland in front of the stream bends
Anton Field: OE *an* = one
(Long) Atton Riggs (1620): OE *ate* = oats
Aple garth (1620): apple orchard
(Nether) Arable close (1620)
Arber Lodge (1623): OE *eorth-burgh* = earthwork. Also arbour = a shady retreat formed by trees or lattice work and creepers
(Brade) Arses (1618): OE *brad* = broad arse (its shape)
Alger-furlang (thirteenth century): a surname
Ashtrees
Ashley and Cookhouse Banks
Aver Acres: 'havers' = oat fields or 'aver' = a beast of burden (H)
(Middle) Awards (1618): land allotted
Axit Hills: OE *aesc-sceat* = ash grove

Backside of the Hill
Backfield
(The) Bad Ground
Bagraw
Baike house close (1585): where there was a bakehouse
Bake House Riggs
Bainshaw Bogs: a 'shaw' is a small wood
Baiting close: 'bait' is dialect for food. A feeding place
Bank Quarter (1620): ME *banke* = slope of the hill
Bank riggs (1620)
Bankey: river or stream bed
(Low) Bankhead
Banke meadow (1614)
The banke nighe the miers (1567): the bank near the boundary
Bar Haugh: bare or barley
Barelaw Flatt (1618): bare or barley
(The) Barley Field Haugh
le Barnes close (1585): enclosure with a barn
(Long) Barracks: OE *bere-wic* = barley farm
Baron House (1472): House belonging to a baron
Bastielands (1614): 'Bast' is the inner bark of a limetree, or other similar fibre, used for mats etc
Bastiford (1585): it has now become 'Bastile' where there was a bastle, a fortified tower? (See above also)
Bastlings
(The Ewe) Batings: feeding place
Batt Bog: in N. Country dialect it is low-lying land liable to flooding
Battons
Battery Field
(West Rig) Baulks: OE *balca* = unploughed division between strips
Bawca's Close: 'Bawber' is a salmon poacher and it is next to Salmon's Field. It could, however, be a personal name
Bean lands (1620)

Beamwham Allotment: OE *beam* = tree, beam; OE *hwamm* = marshy hollow
Beautys Field (1770)
Beer Close: OE *bere* = barley
Beesom Inn Lands: 'beesom' is a broom
Bell shield: shepherd's hut on a hill
Bell lands (1623): either a personal name, or ME *belle* = bell shaped hill
Benacres (1250): bean acres
Benefordacres (1250): bean fields
Beneby Hill: bean fields on a hill
Beneflat (1235)
Benfield dike rigs (1620)
Benny carrs (1620)
Bent: OE *beonet* = bent grass, not good for grazing
Bentlands close (1620)
Berry Hill (1620)
Between Towns: between 'townships' – not in the sense of a modern 'town'
Bierfield: OE *bere* = barley and corn
Bigland Dale: ON *bygg* = barley
Biglichirne (1250)
Binsdale: ('Bing' was a bin for a heap of grain. There was a 'haybing' and a 'cornbing' (H)
Birchy
Birch Hot: OE *birce* = birch; *hot* = holt
Birkbush (1620)
le birkeclose (1585)
Birkhill: birch hill
Birdhopecraig: it could be burn-valley-crag
Birelaw Haugh (1616): possibly a cowshed on a hill
Bitchflatt (1623): land on which birch trees grow
Blackburn Havers (1620): oatfields by the dark stream
Black Chesters high field: 'Black' can refer to the colour of soil, to burnt areas, and so on, but it can also mean bleak
Blackhalfacres (1620): OE *blac* = pale, bleak
Black Briggs (1620): OE *blacc* = black, dark ON *bleikr* = pale

Blacke-hall garth (1585)
Blakforthlande (1498)
Black Bushes
Blackpool
Blackdaw Leazes (1620)
Black middings ford (1620): ME *midding* = midden
Bleasur lands (1624): ON *blesi* = a bare spot on a hillside
Blind wells (1620) OE *blind* = blind, hidden, dark, secret
Blindwell Meadowe Butts (1618)
Blinde well lands (1614): hidden well, spring or stream
Blyndewellflat (1334)
Blother Meadow (1616)
Blowbutts (1618): ME *blot* (pr. 'blote') = bare, desolate
(Near Shovel) Boards (1770): *braid* = broad. Long and thin piece of land
Boat Haugh: riverside mooring
Boden Hill: hill by a curving valley
Bodle Hole Quarter: Bodles, boggles etc. are ghosts, especially when the word is connected with a 'hole', but it could also come from OE *botl* = a building
The Bodles
Bog Close: bog, marsh
Boggle Wood: see 'Bodle'
(The) Boiling Rigs: probably the wind on the crop growing on ridges would make it look as though it were boiling. Another possibility is 'boolin' (bowling played with a rounded water-worn stone, or 'bool' (H). The field is close to the sea, where such stones are easily found
Bonny Rigs: lovely, pretty, fertile
Borewell Close: land where a well was sunk
Botany Bay: A transferred name – a long way to go, or penal servitude to work the field!
(South Pasture) Bottom
Bought Lands: although the land could be 'bought' in one case, there are two other meanings: (a) N. Country dialect for a sheepfold (Scots dialect is 'bought'), (b) 'bout' is the going and coming of a plough along adjacent furrows
Bout Green
Bowt Leeses
Bowell (1485): OE *boga* = a curving boundary
Bowlbank Leaze (1620): perhaps OE *boga* = land with a curving boundary
Bower Shield
Bowling Alley Hills
Braced Law: ON *bra* = brow, edge of the hill
Bradford dike boggs (1624): broad
Bradlaw Hill: OE *brad* = broad hill (*hlaw*)
Brad Roods (1620)
(Sandy) Brae
Braidle nowe (1567): broad meadow on a hill
(Low) Branchy Leazes and High Thick Hirst: wooded meadow
(White) Breads: broad strips with white grass
Breryg (1479): OE *brerig* = overgrown with briars
(Long) Brocks: in this case OE *broc* = brook, but in other cases it can be OE *brocc* = badger
Bronslauemedoue (1479):OE *brun* = brown
Bronslawflate (1479)
Brown Rigg end (pre-seventeenth century)
Broomeletch meadow (1620): OE *brom* = broom
Broom Coat Knowes (1620)
Broomclose (1637)
Bruntone longe crofte (1567): this could be ON *bruni* = a settlement place cleared by burning
Buckley Wellfield: OE *bucc* = buck, meadow
(High) Bucksters (1770): buckthorn – the source of a purgative, charcoal and a green dye
Bukeley Bank East Field: there are still fields where deer frequently appear
Bull Close

Bullion: where the bulls are kept

Burgeam Yate (1585): this is a farm from the burgages, now called Burgham

le burgh ferme (1377)

le burne close (1479)

Burn Knowle roods (1614)

Burn Crook (1577)

(The) Burneyflat

Burrell rigge (1585): either a personal name, or land occupied by burgesses

Burrow garthes (1567) from OE *byrgels* = a burial place

Burrow garthes (1567): burial place? Rabbits?

Bushop Well Meadow (1620): OE *busc-hop* = bush-valley

Butty meadow dales (1602)

Buttterlaw Close: good pasture for cattle

Byer Ways: OE *byre* = cowshed

Bygate: by the gate or road

Cabbin Hill: a cabin is a wooden shelter, store, or watchman's hut (H)

Cadwell: OE *cald* = cold, exposed, bleak

Cairn Field: a cairn is a heap of stones, a burial mound, but cairn, kern and kairn also mean 'harvest'. A kern doll (or kern babby) was made from the last cut of corn

Caldenelburn: cold stream in a valley

Calf Close (1577)

California: A transferred name

Callech (1498): Crow's letch (later Cawledge)

Camp Field: (a) campus = field, (b) prehistoric or other camp, (c) the 'campin' was also the race to finish cutting the corn

Camphill (1620)

Canada: a transferred name

Canonflatte (1479): land owned by the canons

Carr Haugh: OE *carr* = rock

Carredge Field: Old Welsh *carrec* = rock

Carhill Farm: ON *kjarr* = brushwood

Carricares (1624): ME *ker* = swamp

Carrick: here the names are usually to do with rock or marsh, but Carr is also a personal name

Cartington Shield: This appears to be an *ingas* type name

Carter Knowle

Carter Lane (1620)

Castle Field: 'castle' is common and can apply to many fortified structures

Castle close doore (1614)

Castle walls

Catt-heugh doores (1585): OE *catt* = wild cat

Cat Law Meadow

(Wild) Cat crook

Catch Law: it could be like Catcherside, OE *cald, ceald* = cold

Causeway Road Field: ME *cauce, cause* = a paved way (e.g. The Devil's Causeway is a Roman road)

Cavel

Lady Well Cavel: A cabel or cavel was a share of land apportioned by lot or kyevel

Cawdricks: OE *cald* = cold. A 'rick' is a pile, as in a pile of stones as much as in a rick of hay or straw

Cement House Field

Cemetery Field

Chamley gappe (1479): This shows how much spellings change: the modern name is Cheveley, but in 1300 it was Chiveleye = Ceofa's or Cifa's clearing. Many fields take their names from nearby settlement sites

The Chanell Flatt: on the Aln estuary

Chapel Fields: Either belonging to, or near to, a chapel

Cheasehills: OE *ceosol* = gravel land. Care has to be taken with an alternative, OE *cese, cies* = cheese

Cheshill Close (1624)

Chesdon Hawuers (1624)

Cheek (Elsdon Gate) 'check' is the side (H)

Chesters: OE *ceaster, caester* from Latin *castra* = a fortified place

Chesterletch Field (1620)
Chestres Flatte (1620)
Chevy know: perhaps named after the Cheviot (origin unknown)
Chitchacker dick (1585): either OE *cyta* = kite; or church land field in which the church is situated ('Glebe' is used elsewhere)
Church Field
Churchwarden Land (1585)
Cinder Kiln Field: industrial spoil
(West) Clarilaws: it might mean muddy, heavy
Clatteryng forde (1585): OE *clatter* = loose stones; ME *claterand* = clattering, noisy
Clattering Burn (pre-seventeenth century)
Clatteringhouses
Clatterzan Well (1620)
Clay Dale
Clayford butts (1620)
Cleatley Bank field: perhaps OE *claeg* = clay or cleats are fastenings, strengthenings
Cleugh meadow (1614): OE *cloh* = ravine, dell
Cleugh Brae
Clints Walls: Old Danish *klint* = a cliff
Cloddy Lands: possibly a field with clods in it, or heavy and wet, as in 'claggy'. Also OE *clud* = a mass of rock
Closehead: at the 'head' of the enclosed land
Clott Riggs (1618): with clods
Clubshaw flatt (1614): possibly OE *clud* = rock, fields next to a rocky copse, or wood from which clubs or sticks were taken
Coachroad Field
Coaly Way: a lane to the old coal pits
(The) Coal Bank
Coat Close (1620): Normally *cot* = cottage, but in the Northumberland use of this there is also a sense of 'on the borders of'
Coat flatt stinting (1620) A stinting is an area for grazing animals, or a share; see 'gate
Cotley Hill Banks

Coatlands
Coatland Rood (1620): this is on the coast
Cobden High Pasture: OE *cob* = summit OE *dun* = hill
Cockle Ridge: OE *coccul* = a weed that grows in cornfields (corn- cockle), or OE *cocchyll* = woodcock
Colchas (1623): OE *col-shaw* = charcoal burners' wood
Coldgate Haugh Banks
Cole halfe acres (1620)
Colley Law Field: OE *col* = coal, charcoal
Colsaw (1250)
Colt Land Balks: land on which young horses are kept
Colwellhill: OE *col* = cool stream or spring
The Comb Hill: OE *cumb(a)* = valley
Common Flatt foard (1620): ford at the common land
Common Flatt Fields: the common furlongs and balks in the open-field system
Comon Balkes (1618)
Connrigs (1577): ME *coni(n)ger* = rabbit warrens
Connygarth (1585)
Copt Hill: OE *copp* = a hilltop
Corby's Hill: a 'corby' is a crow
Corbys Bounds
Cork Leach Ley: as part of this refers to a pasture or meadow made by a letch (marshy stream), 'cork' could be corf = a basket made of hazel rods (corf-rods)
Corn Field
Corney Horner: the corn field on a corner (in this case a right-angled bend in the lane), but 'corney' can also mean drunk
Corner Haugh: the field is in the bend of a river
Cottage Close
Cottage Field
Cottenshope: (there is a Cotteneshopp in 1230), probably Cotten's or Cotta's valley
Cotyards demeane (1614): OE *cot* = cottage, hut

Couchel: OE *cwice* = couch grass
Coultherd Meadow: OE *culter* = coulter
Council House Field
Cow'd Oak Field: cold?
Cow Grass
Cowey Sike Middle Field: OE *cu(ge) haeg-sic*: land enclosed near a sike to keep cattle in
Cowstand (1620)
(Muckle) Craigg: a big crag
(Lower) Crag Field
Craggies Corner (1770)
Crankesses: OE *cran* = crane, heron
Craustrige (1479): land by a cross, or lying across
(West) Crawlaw: OE *crawe* = crow; west crow hill
Crawley Whins: crow-meadow where gorse grows
Crendale: OE *cran* = crane, heron
Crestwell Bog Field (as in Kereswell, 1234): a spring where watercress grew
Cricket Field
Crinkey Bush
Croftes Deales (1620): a croft is a small enclosure for tillage or pasture
Croft Lowpiece
Crookitt roods (1614): ON = *krokr*; ME *crok* = crook or bend. There are many examples by rivers and streams
(Common Burn) Crooks
Crookletch Pastures (1620)
Crop Close
Cross Letch Riggs (1620): in some cases 'cross' refers to the way fields run into each other, or the way a lane crosses a field, but in others there is the possibility that there was a cross in or near the field; OE *cruc* = a cross, but likely to be the same as 'crook'
Cross Hill (pre-seventeenth century)
Cross flatt (1577)
The Cross Loan (1577)
le Crosse (1479)

The Cross Field
Cruckses (1618)
Crum Roods: OE *crum* = crooked, but in dialect 'crum' can mean a cow or 'plump' when applied to good quality produce (H)
Crum Steeles (1614)
Crummy hawuers (1614)
Crysedale (1250): ON *griss* = a young pig, so this could be the land where piglets were reared
Cubstocks Allotment
Cuttle Quarter (1621): ME *cut* = a channel. The field would be drained by deep furrows
Cutley Hill
Cutts Hill Croft
Cufleys medow (1620): not clear, but (a) a 'cuff' is a simpleton (H), (b) OE *cu* = cow
(East) Cufleyes

Daisey Field
Dale: OE *dal* = share of the common field; OE *dael* = valley. Deal, dale and dole all mean 'divide'
(The) Dealls Pasture
Dam Close
Damheads (1618)
Dana Riggs (1623)
Daneley medow (1624): perhaps 'dean' – thus a meadow by a valley
(Jenny's) Darg: a darg, dargue, dorg is a day's work (in an Elsdon Terrier it says '9 dorgs of meadow. 13 rigs being 4 dorg'). The amount varies – it could be a day's ploughing or mowing
Darn Burn Close
Dean: OE *denu* = valley
Deans mill (1577)
Dear Boughts: this sounds as though it cost too much, but note that 'bought' can be dialect for 'sheepfold' and ME *bouzt* = bend, turn
Deer Park

Delf Rigs: OE (*ge*) *delf* = quarry and locally 'delfs' are small pits
Delf lands (1624)
Demain Close: demesne land. The word sometimes becomes 'mains'
Demesne Dyke (1620)
Diesel Tank Field
Dike: a dike or dyke can be an embankment against flooding, a ditch, or a natural ridge. It appears frequently in early maps and boundary surveys as a ditch separating different areas. The upturn from the ditch could be planted with a quickset hedge. OE *dic* = ditch, drain. Its common meaning is a wall
Dimney Well Field
Dipper Field: this field dips; but there is the Dipper bird, too
Disputed Land
Divshey dycke (1585)
(Green) Dodd: ME *dodde* = a rounded hill
Doe Path
Doffenbee Flatt ?
Dog Holes: possibly dog's tail grass or dock growing at the old kiln
Dolakelawe (1295)
Donking Rigg Corn Ground: a donkindale is a mist rising in hollows at evening, dunk, damp, dank (H); it is also a personal name
Doupster
(Dirty) Doup: ON *daup*. It means 'buttocks' – and it is not unusual to find 'arse' in a place name too. Doup is also the bottom or end of anything, a rounded cavity, a hollow bottom.
Dovecote Close: the contents of the dove cotes were valuable additions to a fairly sparce winter diet, and played an important part in domestic economy
Dove Cote Close (1612)
Dowkerhalgh (1472): ME *dowker* = diver
Dowkerhalgh (1472)
(Long) Draughts(1620): long strips of land

Draw Kiln Field
Dudlees: ME *dodde* = meadow by the hill
Dueldrigge (1479): presumably the place where duels were fought? Or 'double'?
Duelling Haugh
Dumhead farm: dam or OE *dumpel* = in a depression
Dungell Hoopes(1614): OE *dung*; land that was manured or where dung was stored
Dunsdole: hill
Dunsheld: OE *dun* = hill; OE *helde* = slope. Steeply sloping hill
Dunshawe pasture: OE *dun-hoh* or shaw. Pasture by a wooded hill
Dunnsheugh Wood: wood on a steep hill
Dunslaw Flatt: field by the hill
Dunstone Links: OE *dun-stan* is a rocky hill
Durte poote butts (1614): shares in the common field at the place where there were dirt pits
Durtrees: This may have the same origin as the above
(Potts) durtrees

Earls meadow Ancient Land
Earsland roods (1624): OE *ears* = buttocks, rounded hill
le east close (1585)
Easter Awards (1618): land allocated on the east
Easterfield (1620)
Easter Seaton Close (1614): eastern
Ebrockes (1250)
Eccleshalghforth (1471): the ford at the haugh by the church
Edynwell (1394): St Aidan's well
Eels Flat: OE *ea* = river, stream. Land by a river
Eight Acre Field
Element Hole
Elishaw Mill: OE *el* = a small island
Ellenfield: OE *elle(r)* = elder tree
Ellerwell Field: elder tree

Elley Green: elder tree

(Great and Small) Ell Field: in addition to the above, 'ell' can refer to the shape of a field and to a measurement

(Long) Elmes (1577): land by the elm trees

Elstrother Yate (1620): OE *ellern* = elder tree. 'Strother' is land overgrown with brushwood

Elyhaughe (1585): OE *el*; ME *ele* = a small island in the river (Coquet)

Endemyre (1498): land at the boundary

(Haver) Ends: OE *ende* = outlying part

Engine Banks: connected with mining

Ermet-fall meadow (1470): ME *ermite* = hermit

Errington's Haugh: place, not person

Esellfoord Lands (1620): ON *eski* = ash tree; OE *aesc* = ash tree. Lands near the ford where the ash trees grow

Espette fourde (1620): OE *aespe* = aspen tree

Estcrosflat (1334)

Est Noyke (1585): east

Estraustrige (1470)

Evenstones: smooth, level

Ewe Hill

Eye Rigs: 'Eye' can mean any of these things: (a) hole in a pick or grindstone, (b) an opening at a watermill, (c) a discharge hole in a limekiln, (d) the mouth of a well, (e) land by water, yet in this case it could be from MONEYE (see below)

Face of Park Hill Road: facing

Fairney Hills: OE *fearnig-ley* = ferny grove or clearing

Fairnley

Fairy Knows

Faldar dike (1567): OE *fald* = fold

Fallowlees Moor: OE *fealu* = fallow, pale brown

Faltemere (1480): OE (*ge*) *fall* = clearing; OE *falh* = ploughed land; OE *fald* = fold, enclosure

Fanney Leazes (1623): meadow, or fern-covered

Far Crookes (1624): bends

Far Little Field

Farney Knowle: fern-covered hill-top

(the) farry rodes (1585): either far or ON *fagr* = pleasant roods

Fartimerethorne (1480): thorns growing either at the far pool or boundary? See Faltemere, which is the same place

(The Fastlings): Presumably where the animals were held fast (like OE *falding* = the folding of animals)

Fatting Pasture Faugh: OE *falding* = where the animals became fat

Faugh: OE *falh* = ploughed, fallow land; OE (*ge*) *fall* = clearing

Fawdon meadows Field: OE *falh* = fallow

Feather Blow: light, sandy soil by the sea

Featherblow grounds (1620)

Fether Blew (1567)

Featherwood Common

Fell Bulls Close

Fell Close

Fenham Butts (1618): settlement by the marshy land

le fense (1377): boundary fence

Fertles dyke (1585)

Fislebee Pasture: fissle, fissel = to move restlessly with a gentle crackling noise (H). This could be the movement of beasts on straw

Firth Moor: perhaps 'furze' in this case

Five Corner Field

Five Rigg Field

Flase Lands (1614)

Flase Meadow (1614): ME *flasshe* = swamp

Flatt Fell

Flatt of Lands (1618): a division of the common field, but ME *flat* is also a level piece of ground

the Flatt Meadow (1614)

Fleet Gate Close: OE *fleot* = stream; 'gate' in this case is a road

Fleets

Fletys (thirteenth century)

Floor Crooks: flat land at the bend of a stream. It could mean a threshing floor in other cases

Flowery Field

Foggie lands (1618): OE *fogga* = long grass. 'Fog' is dialect for coarse, rank grass

Foggy lands

Foggyleas style (1580)

(The Night) Folds: where beasts were herded at night

Folley Close: OE *folla* = foal

Fool pool

Footpath Field

Ford

(The) Fore Close

Foreshield Grains: in front

Foreside of the Hill

Forth Gate

Forty Pound Field

(The Willow) Foss: Latin *fossa* = a ditch

foule brigges (1580)

Foule Meere Riggs (1616): OE *ful* = foul, dirty

Foul Haggars

(The) Four Demain Riggs: belonging to the demesne lands

Foure lands (1614): OE *furh* = furrow, trench; OE *fore* = land in front

Fou(r)mert Letch: a watercourse through boggy land

Fountain Close

Fox Holes

Franklin Well Field

The Frears Stone (1567): it adds: 'a greate longe stone, hewen on ye one ende'

(Small) Freeholders Allotment

Freshmire Bank (1620): ON *mere* = a pool into which 'freshets', or streams, flow

(Park) Frith: OE *frith* = woodland

(The) Front Field

Fryar Pool: belonging to a friar, or where there were young fish?

Fryer Park Meadow (1620): the site is near Brainshaugh Priory

Fulrig (1470) Furlongs: foul, dirty

Gaile Close (1620)

Gallas Knowle (1620): OE *galla* = barren, wet land

Gallowfield Haugh: OE *galga* = gallows

Galls: There is just a possibility that this name has a different origin from the above, for a gall is an excrescence formed on trees in response to the presence of insect larvae, and this swelling in the case of oak-galls is used in the manufacture of ink, tannin, dyes and medicine (Oxford English Dictionary)

Garbutts (1618): OE *gara* = gore, a triangular piece of ground

Garden Close

Garretshields: (French *garite* is a place of refuge)

(The Demain) Garth: ON *garth* = enclosure, yard

Gate: (a) A gate is a stint, or right of pasture; 1 stint = a cattle gate; 5 sheep = 1 stint. A cowgate is a stint. A coble gate is the right of fishing (H), (b) OE *geat* = gate, (c) ON *gata* = a road, (d) OE = a goat. These alternatives show how difficult it is to be certain about a meaning, especially because many interpretations can be equally likely

Gateness Hill: a 'ness' is a headland

Gawland Quarter (1620)

Gewle lands (1618): see 'Gowland'

Ghirletch Meadow (1620): could it be from OE *gear* = a device for catching fish?

Gilders (1620): ON *gildri* = a trap or snare

Gills Hole Field: as this is by a 'cleugh' and a river, 'gill' is used in the sense of ravine. The 'hollow' reinforces this

Gilstrother: stream, rather than ravine

Gin Field: where there is a 'gin' = a machine

Girsonfield: girsonne = a fee paid on renewal of a lease. Also Gressom, Gryssom (H)
Girstlerley (1620) ME *agist* = rented pasture
(The) Glander Close
Glebe Lands: belonging to the church
Glebeland Uplands (1621)
Glegarims Corner (1770) ?
Glittering Stones Field: glidder, glitter = loose, rolling stone, scree
Gibbones dike nooke (1567)
(The Calf) Goat: OE *gota* = a water-course. A 'gote' locally is a stream approaching the sea through sand or slake (H), but this one is inland and marked 'a drain' on the 1760 map
Gonuldes Cross (1295)
The Good Ground
Goodly Crawlaw: fertile land
Goose holm: OE *gos* = goose. The holm is a low-lying piece of ground by a stream
Gorbut (1616): OE *gara* = a triangular piece of ground
Governor's Close
Gowkstone Field: a 'gowk' is a cuckoo
Gowland Croke poole (1618): it might be 'golden' or 'marigold' – thus a pool at the bend of a stream where the golden flowers grow.
Graindge borne (1567): a grange is the outlying part of an abbey
(High) Granary Close
Grange Path (1577)
(Hills of) Grain
Gravel Bed
Graye-stone well-heid (1580): a marker stone at the place where the spring begins
Greanshields (1620): ME *schele* = hut or shed on the green field
Granshels (1620)
Great Corn Hill
Great Field
Grenewell flat (1334)
Greeneletch (1618)

Green garth (1580)
Greenside Latch (pre-seventeenth century): grassland
Greenrigg
Green Orchard (1620)
(Town) Green
Gresse-yarde (1580): ON *gres* = grass; OE *gaers, graes*
Grey Stone Close: a boundary marker
Griffin buttes (1614)
Grimping Haugh: possibly French *grimper* = to clamber up
Grind Lands (1620)
Grysgarth (1498): either grass or a piglet

(The) Hacks (1770)
Hagley Rigs Haining: ME *hacket* = a piece of cleared ground; or OE *haca* = thorn tree
Hagg: OE *haga* = fence, fenced enclosure. A peat hag is a projecting mass of peat
Hagberry Field: hag-, heg-, hacker-, hackberry is the fruit of the bird cherry, Prunus padus (H)
Hagley Riggs: OE *haga* = fenced enclosure
Haining: OE *(ge)haeg* = hay, enclosed land, meadow OE *haegen*; ON *hegning* = an enclosure or grove; ME *hay* = forest fenced off for hunting
Hainging Hawuers (1620): oat lands that have been enclosed
Haisand (1567): now Hazon. The end of the fenced enclosure
Halfacre butts (1620)
Half Acre Close
Half Crown: most likely to be its shape
Halley Well butts (1614): OE *halig* = holy
Hall-stede (1472): OE *hall* = property of the lord of the manor. Sometimes it can be from OE *halh* = nook, land in the bend of a river (haugh)
(The) Hall West Pasture
Haltram bright riggs (1623)
Hamelspeth (1585): Usually 'hemmel' is a

cattle shed. OE *hamel* also means bare, 'peth' is path
Hammund(?) gate butts (1620)
Hampeth Fourde (1567): homestead-path-ford
(The) Handkerchief: a fanciful name for a small piece of land
Hanging balke hawuers (1614): OE *hangende* = land on a steep slope
Hanging Dales
Hangwell Quarter
Hangmanacre (1485): land allocated to the hangman
Harbour Dikes
Hard Luck
Hardon (1620)
Hard Rig: land difficult to manage
Harelaw (1620)
Hareside roodes (1620)
Haresyde (1620)
Haris flatt (1620): hare
Harker Snypes (1620): it might be ON *haukr* = hawk; OE *snaep* = marshy land. It is interesting that the snipe is a bird that lives in marshy land
Harlott Sheete (1615): in the open-field system a 'sheth' was a group of parallel strips of ploughed land which joined a similar group at right angles. 'Harlott' could be 'hare'
Harrowgates: this could be an implement. There is also OE *haer* = a pile of stones, a cairn
Harow Hill (1620): OE *hearg* = a pagan sacred grove
Harlow Hill Pasture
Harpers Heughe (1577): this could be a fanciful name for a triangular shape, or the land belonged to a harper
Hathery Batt (1621): OE *haeddre* = heather
Hathery Close
(Long) Hatt? Perhaps a mis-reading of 'batt'
Haughey Close: a 'haugh' is land by a river, usually enclosed in a bend
Haugh Field
Haughslopp butts (1620): 'slopp' is either OE *sloh* = mire or *slah* = sloe
The Havers: ON *hafri* = oats
Hawkses
Hawkeys Close: (a) hawks, (b) a 'hawk' is an implement for unloading and spreading manure, (c) a Hawkie is a white-faced or pet cow (H)
Hawsop Well Field (also Hassop and Haslop in the same area): there is a local surname, Heslop; ON *hesli* = hazel
(Short) Hawthornes (1614)
Haxton Brow
Hay Riggs: either land for hay, or land enclosed by hedges (OE *(ge)haeg*)
Headland Riggs (1620): OE *heafod* = headland, head, end of a ridge, source of a river, upper-end. Many 'headlands' are the places where the plough turns
Headlands
Headlawe (1580)
Heads Know
Heal Swans: Swan is a personal name. 'Heal' could be high land by the bends of a stream
Heapland Crooks
Heather Moors
Heckley House Field: OE *hea-clif* = high cliffs; or heather cliff
Heckside OE *haec* = hatch, a small sluice gate suspended from a pole
Heddon Bank: it might be high hill (*dun*), or the high bank of a valley (*denu*)
Hedge-croft (1567)
Hedgelaw Flatt (1621): hedge hill
Hedland: see 'Headlands'
Hedow Bush (1620)
Hedshope: head of the valley
Heelywell Crookflatt (1621): either holy well or *holh-wella* = spring in a low hollow
Heighfer Lands: 'hefe' and 'heaf' are regular

pastures for sheep
(Low) Heilds (1614): OE *helde* = slope
Heild Hawuers (1618): sloping land, good for growing oats
Heldon buttes (1614): could be a hill-slope
(South) Helfers: see 'Heighfer Lands'
Helm Field: place name, a roofed shelter for cattle
Hemmell (and Hemel) Field: with a cattle shed
Hending Riggs (1623): OE *eng* = oats
Hen Law (1616): hen or water-hen
Hepies way roods (1642): a roadway
Hepster Butts: hips
Hepstrother: land covered with brushwood, where rose hips grew
Herdlaw: OE *hirde* = herdsman's hill
Hesley butts (1620): hazely
Hesperlaw (1620): OE *aespe* = aspen tree (hill)
Hether Side (1614)
(lez) hevedlandes de Brouneslawflatte (1479): the oatlands of Brown's hill flat
(West) Hidgley: meadow enclosed with a hedge
Hield Field: field on a slope
Highlands Green
High of Chesdon (1620): high part of the gravelly valley
High Rigg Intake: high land enclosed for cultivation
High Seas: where the wind moves over grass or cereals
High Whittridgefield (1621)
(The) Hill Burn
Hindup
Hinds Shield: deer
Hingeyshea top
Hinging Well Strand: OE *hangende* = steep, overhanging
Hingston Lee: the lee of a hill slope
Hipsbum Moor: rosehips
Hirds Hill (1580)

Hitchcroft (1620): ME *hiche* = a croft enclosed with hurdles
Hither Holme: a nearby water meadow
(Low) Hive Acres
Hoa Dales: OE *hoh* = heel, projecting ridge
Hodden Tippet common meadow (1614): OE *denu* = valley
(Wide) hoe Haugh: OE *hol* = hole, hollow
(The) Hoeings
(Small) Holdings
Hole Acre: OE *hol* = hole, hollow
(West Tod) Hole
Holes Croft
Holemeadow and lands (1620)
le Hole Croft (1580)
(The) Holm: ON *holm* = island, water meadow, riverside land
Holmersbank (1470)
(Low) Holmes Field
Hollow Burn Close
Holy burn close: holy or holly
Home Pasture
Honey Spott: where there were beehives
Horners Field
(The) Homes (seventeenth century)
Horns Know (1770)
Hoods Knowe (1620)
Hope foot: valley
(The) Hope (1295)
Hoppins: OE *hoppit* = a small enclosure
Horse Close
Horse Reins: horse pasture at the boundary
Horslawpule (1470): horse hill pool
Horslaws (1620)
Horsley: horse pasture
Horsehoe: usually the shape of a field, although the present shape may not fit the description
Hot Bog: OE *hot* = holt, a small wood or thicket
Hotlaw Brow (1620); thicket on a hill
House Pasture
(the) houghe (1580): same as 'haugh'

Howdeen: in the following examples the origins may be: (a) OE *hoh* = a projecting ridge (heugh), (b) OE *hol, holh* = hole, hollow
Howdens (1580)
Howgate Waste
Howlekilne water (1620)
Howle meadowe (1614)
Howle Wilbub butts
(The) Howme
Howmer latch butts (1620)
Howpe Butts (1618): OE *hop* = valley
Hudesrodes (1470): it seems that Hud is a personal name although it could mean a headland
Huds Head
Hudspeth Common
Hudletch meadowe (1614)
(The Great) Hughe, or Cragg, called the West Heughe (1577): OE *hoh*
Humbles Knowl (1621): OE *hamel* = bare
Hungerfull lases (1620): OE *hungor* = hunger, poor (of land)
Hungerford Law
Hungery butts (1614): OE *hungrig* = poor, infertile
Hungerup West Field
Hungry House Farm
Hungreknoll (1471)
Hunters Close

(Long) Imberley (1623) ?
Improvement: taken in from waste
Inlands (Scotch Arms Inland and Lonning House): OE *inland* = land near the homestead
Intakes: ON *intak* = taken in from waste
Iron (High Shaw and Iron House): OE *hyrne* = land in a corner, but Ion, Ions are local surnames
Irish Dales Izlands (1620): OE *ea-lands* = field either surrounded by water or by other fields

Jobs Fields: biblical; a borrowing? Could the fields have needed a great deal of patience?
Jockeys Field
Juniper Haugh Waste
Junction Field: a railway, in this case

Kaile yard hawuers, butts, meadow (1614): cabbage or cow hill
Katakre (1334): OE *catt* = wild cat
Kakney Leaze (1620)
Karse Burns Meadow (1620): probably watercress
Kayhill close (1526): OE *cu* = cow
Kyhill-close
Keenes acres (1624)
Kelly burn: possibly ON *kelda* = a spring, but it could be a personal name
Kerse well (1585): watercress spring
Kilnflate (1479): OE *cyln* = kiln (limestone)
Kilne flatt (1621)
Kiln Field
Kiln Croft (1637)
Kilterton riggs (1620)
Kiltert Riggs (1620)
King Parvo
Kings Cross a Bank: OE *cyning* = king, and a surname
(the) King's gates (pre-seventeenth century)
Kingshaw Haugh
Kingslaw: king's hill
Kip Hill
Kippitt Law (1620)
Kirkcroft (1620): church
Kirkhill Pasture
Kirkways
Kim's Dale
Kirspewell Hawghe (1567): watercress spring haugh
Kits Know
Kitwell Lough (1620): OE *cyta* = kite
(The) Kitty Catts (1770)
Klondyke: a transferred name

(Fairney Brae) Knab (1770): OE *cnaepp* = hillock, the slopes covered with ferns
Knightside Ancient Land
Knotforth foord (1620)
(Heads) Knowe: OE *cnoll* = hillock
Korhilles (1479): corn hills or crow hills
Kyelstone forde (1585): cattle (hill) stone ford

Labour in Vain: a comment on the condition of the land
Ladyman Field
Lady Lands (1620): the rent may have been due on Lady Day, 25 March, or the lands could have been dedicated to Mary for maintaining a chapel or shrine
Laffer Lands (1614): it could be OE *leah* = woodland; clearing, open land
Laine (1621): by a lane
Laingshill Low Pasture: either long hill-pasture or from 'leyne'
Lainsh Crosse medow (1585): ME *leyne* = arable land
(Ewe) Lains: OE *laning* = lane, tracks
Lambe meadow (1585)
Lamehill lands: ME *leme* = an artificial watercourse
Lands Stinton (1623): stinted land, shared out
Lancelot's Haugh: probably a fanciful name
(Old Pasture) Land: a land is a strip, a division of the old field
Landshott End Ancient Land
Lane Field
Lang Carton parcel of Bilton Oxpasture (1620)
Langelands (1250): OE *lang* = long, long pieces
Lang Hewgh (1585): of land, long strips
Lanshaw Rig: long copse, unless it is derived from ME *leyne*
Lases: meadow
Laverick Law: OE *lawerce* = skylark
(le) lawe (1585): OE *hlaw* = hill, but ON *lagr* (pronounced 'law') means 'low lands', where many arable strips lay
(East) Law Moor: hill or mound
Lawns House Field: ME *launde* = woodland glade
Lazy Hill: unproductive, or from 'leazes'. *Lazy beds* was also a method of planting potatoes
(The) Lea Field
Ley: field
Leazes: fields
The Leys (1250): could be OE *laege* = fallow
Leach Pool Haugh: either the pool had leaches in it or a 'letch' ran into the pool
Lead Haugh: OE *(ge)lad* = watercourse
Lead Hill Farm
(The) Leinings: ME *leyne* = an enclosure of arable land
Leme lands (1620): ME *leme* = an artificial watercourse
Letch: a long narrow swamp in which water moves slowly among rushes and grass (H)
(The) Levels
Lidget Moor: OE *hlid-geat*; ME *lid-gate*, dialect *liggat(e)*, *ligget*, is a swing gate that prevents cattle from straying
Ligger
Light Pipe: a 'pipe' is a thin seam of coal (H) or a watercourse
Lilly Close: Lilly-low, Lilly-lee = a grassy slope (H)
Lime Kiln Holes: quarry
Limestone Quarry Kiln
Limey Crook: ME *lem* = land with artificial watercourses
Lindley Haugh
Lin Croft: linn = water, pool
Linkeyletch (1618): OE *hlinc* = ridge, terrace on a slope (like lynchet)
Linkey Law: terraced hillside
Link Closes: 'Links' on the coast are sand dunes
Linnen Cleugh
Lintburn Close: OE *lind* = limetree; OE *lin* = flax

(Neither) lintland butts (1620): land growing flax
Lintle Hill
Lipper Edge: could be OE *hliep-geat*, a fence that deer could leap, but which kept in cattle
Little Field
Little Scotland (1585): the name of a stream; ME 'scot' was a tax on land, so this stream could be the boundary of such land
Loan
Loaning Field: OE *laning* = a lane
Lonning
Long Close
Longfloor haver (1624)
Long Knotforth (1620): ON *knottr* = hillock
Long Mary Knows (1618)
Longstrey meades (1621)
le Lons (1479): the lane, road
le lonynghed (1479): top of the road
Look Out West Close
Lords Sixteenth (out of 16 strips)
(Little) Lough (1622): OE *loc* = pen, fold
Lough Rig: near a pool
Lowhoop (1620): a bend on the river below the town
Lowndlands (1620): ON *hundr* = a small wood
(Croft) Lowpiece
Lumsdon Common Land: OE *lumm* = a pool, *denu* = valley
Lurpot (1620): it might be from OE *lort* = dirt; ME *pot* = deep hole, pit
Lutcheside Close ?
Lyherbes leases (1567): OE *laeg* = untilled
Lyllie butts (1614): grassy slope
Lyme pit butts (1614): land near the lime pits

Maddy rig (1702): madder – a plant from which dye was extracted
Maddys Pasture
Madge Haugh (1616)
Mains: demesne lands

Manside Close: 'man' is dialect for mound, a pile of stones (In 'The Poind and his Man' at Shaftoe Crags, the 'Man' is a large burial mound. There is also Manside Cross, a prehistoric earthwork)
Mansion House Garden and Garths
Marche bourne (1585)
Marche Hill (1585): OE *mearc* = boundary
Marks Hill
(Great) Marden: a stream valley that is a boundary
(The) Mare Close: OE *mere* = mare, but it is more usually from OE *(ge)maerc* = a boundary
Marley Mearc Butts (1623): land at the edge of the boundary of the clay
Mart Close: a 'mart' in North Country dialect is a fat beast, but H points out that it is also a bullock that was bought by two or more people to share (in this sense now obsolete). Generally it is now a livestock market
Mary Medows (1620): The hospital was 'Maudeleys'. St Mary Magdalen
Maudeleynwell (1393)
Mawdelyn Croft (1471)
(The) Maynes (1620): ME *main*, *mesne* = demesne land
Mickle daile Howuen (1620): OE *micel* = great (dales are strips of land. Oats)
Mickledayl
Maynflatt (1479)
Meare letche (1585): boundary along the letch
Meadops Riding: probably a clearing for a meadow
Meadow dales (1621)
Meeting House Fields
Merlpottes (1479): pits from which clay is dug
Messgates: moss-gates are stints of pasture on mossy ground. OE *mos* = moss, lichen, bog, swamp

(le) Messeway (1479): lane through swamp
Merstialawes (1290) ?
Mickle daile Howuen (1620)
Middilham flat (1498): OE *middle* = middle; *ham* = settlement
Midge Haugh: OE *micge* = liquid drainings from manure. Both places are on streams. The alternative meaning would be OE *mycg* = gnats
Midge Park
(Ten) Mile Stone Close
Mill Close
Miller's Intake: OE *mylne* = mill
(The) Mill meadow (1577)
Milne hallands: OE *mylne*. The mill on the land belonging to the hall
Milnrig (1479)
The Mill Willyway (1577): OE *wilig* = willows. The willow-lined track that led to the mill
Mirehouse Allotment: 'mire' in this is a boundary
Mirriches: OE *myrr* = swampy; OE *myrige* = pleasant
(High) Moat: ME *mote* = a mound
(Well) Monay: land subject to a special money payment (it can become 'Monday')
Monkridge: OE *munuc* = monk
Mooly fludds (1620)
Moor OE *mor* = barren, waste land
Moralees: OE *moriga(n)-leag* = swamp clearing. This has become a local surname
Morelaw (1479): waste land on the hill
Morrick Riggs (1616): at Chatton. Farm on the waste ground
(Long) Morrifur lands (1614): OE *mor* = fen, waste
(Long) Morthur lands (1620)
Morye butt (1580)
Moscroppe (1577): land covered with moss
Mosscrop Close: also a local surname
Mossfield
Mosvcrokes (1250): moss-covered land on the bends of a river
Mouldshaugh: moles, perhaps, or ON *mold* = top of the head; Welsh *moel* = bare hill. The site is on a small hill looking over the haugh to the river
Mowdie floores (1620): muddy floors between the ridges
Mucke slopp Buttes (1620)
Muckle Craigg: a large crag
Murridge: ON *myrr* = mire, bog

Narrow Croft
Neatherlands (1618): OE *neothera* = nether, lower, lands furthest away
Nether Close
Nether Cruiksses (1620)
Netheredlange (1334)
Neither Shield-dyke (1567)
Nell Walls
Nent Force: a waterfall on the river Nent
Newfoord Peece (1620)
New Rift (1620): newly taken into cultivation
Newlands
New Ryste (1620)
Newstead In: OE *stede* = site, farm. Land by the new farm
(The Niddles): The second part could be 'dales'. ON *naddr* = point, so it would then mean strips of land going into a point. However, this is only guesswork
Nightfield
Night Fold: the place where animals are secured at night
Ninch Knows
Nine lands (1620): nine strips of land
Ninewell Heads
(the) Nook Field: ME *nok* = nook
(the) Noyke of Broxfeld Medowe (1580)
Northlough Close (1620)
Notley Know: hill-top where nuts grow
Nun Flatt
Nunne Close (1567)

Nursery Field

Oak Hill Close
Oaks Scroggs (pre-seventeenth century): oak scrubland
(The) Oates (1620): land on which oats are grown
Obelisk Field: the obelisk celebrates Davison's friendship with Nelson
Okletch flatte (1620): oak-letch
(The) Old Pasture
Onstead (Harlow Hill Onstead Garths): garths (yards, enclosures) on the site
Orchard Field
Orchard Layning (1618): the track around the Orchard field
le orcherd (1377)
The Orcheyarde dyke (1567): orchard wall
Ottercops: otter hills
(The) Out or West Pasture: furthest away
Out Peece (1620)
Outler Hill: either furthest out, or outlet = pasture adjoining the winter cattle sheds
Overacres: a field or parcel of strips lying at right angles to another, and crossing its boundaries
Overschotlaubankes (1479)
Overthwart Riggs (1620)
Oxincruke (1290): ox pasture by the stream bend
Ox Pasture
Oxe Pasture (1621)

Paddowell (1577): OE *pearroc* = paddock, grass enclosure
Page Croft Field
Pan Close (1620): a depression in a flat field
Pandon Heads
(le) Park (1377): OF and ME *park* = land enclosed for hunting or for a pleasure garden
Parlour Field
Parmisford (pre-seventeenth century)
Pasture: ME *pasture* = grazing land

Paunchfords
Pavyne (1577)
Paynes Ford (1620)
Peacock House East Allotment
(The) Pea Field
Peasecod Close
Peaslet
(The) Peat Holes: ME *pete*. Holes from which peat has been dug
(Rochester) Peel: a defensive tower
Peighin: OE *piced* = land coming to a point; ME *pichel* = a small piece of land
Penny Laws Near Field: penny rent? a hoard of coins? small pieces of land?
Pepott (1616): (a) plants – peppermint, field pepperwort, (b) peppercorn rent
Pepper Moor
Perie acres (1614): OE *peru* = pear; *pirige* = pears
Peter lands (1620)
Petty Knows: possibly small hills
Pewitt Haugh
Pheasant Field
Pickle (Tylhirn? House Pickle): ME *pightel*, *pichel* = a small piece of land
Piggs Inn Close: Pigg is a local surname. It could also be from *pightel*
Pigs Field:
Pilchesse lands (1614): ME *piled* = barked; OE *aesc* = ash tree; sugar from ash tree bark
Pilfer Lands: either OE *pyll* or ME *piled*
Pilmoor: OE *pyll* = pool (Welsh is *pwll*)
Pinder Hill: OE *pundere* = land allocated to the keeper of the parish pound, or the pound itself
Pinfold: OE *pynd-fald* = a pound for stray animals
(High) Pingle: ME *pingel* = a small piece of land
(Light) Pipe: 'pipe' is a thin seam of coal or watercourse
Piper Bank
Piperside butts (1620)
(The) Pit Field: coal pit

Pit Rig
(Far) Plains: flat meadowland, an open tract of land
Plantation
Planting Field
Plea Haugh: possibly the subject of a legal dispute
(Mount) Pleasant
Plondon Park (1614)
Plumbtree Hill (1620)
(The West) Plumbtree
Pond Quarter
Poned Lands
Ponitons Close
Pony Field
Poote lands (1614): Probably pits, but 'poots' are unfledged birds – young gamebirds. It can also mean little, insignificant
Pootes wayst (1614)
Possetts leche (1567)
Pottle Prick: ME *potte* = land covered with holes, pricked in; or ME *pightel* = a small piece of land
Pounderclose (1498) where animals were penned
(High) Pows: *poll* = hole, pit, pond which in ME came to mean stream. These 'Pows' are by the River Coquet. (There is a Powburn on the River Breamish.)
(High) Pows
Priest Deane (1620): OE *preost* = priest
Priests Meadow
Prior Park Field
Priors
Preost dean ford (pre-seventeenth century): priest-valley-ford
(The) Pry Closes: ME *pry* = carex grass. In Northumberland 'pry' is the name given to several grasses, and it also names the only part of old grass that sheep will eat – the bottom
Pump Field
Pygles Close (1585): ME *pightel* = small enclosure

(The) Pynd-fold (1567): Animal pound

Quakers Close
Quality Walk: where people 'of quality' walk
Quarel flat (1334): ME *quarrelle* = quarry
Quarrell hedd (1585)
the Quarrells (1585)
Quarrel Whins
(Old) Quarter: quarter of a township
Querredlanges (1334)

Rabbit Hill Pasture
Rails mire and Rayle Myre (1620): ON *ra* = boundary, also 'mire'
Rankstaneflat (1334) A 'flatt' marked by a row of stones?
Rashy Pool Quarter: most probably 'rushy', but a 'rash' is a narrow piece of unploughed land
(The) Rasher Field: rash, rasher, resher = a rush
Rashercap: the reed bunting
Ratcheugh Wood
Ratchwood (pre-17th century)
Rattenrow: OF *raton*, ME *ratton* = rat. Probably rat-infested
Ravens Crag
Raw Ancient Land: OE *raw* = a row, of houses, trees etc.
Rawgreen West Allotment
Rawthurnolech (1498): a row of thorn bushes or trees along the letch
Ray Well (pre-seventeenth century): spring in a nook
Read Ford (1577)
Read forde (1585): OE *hreod* = reed
Read Close Acres
Reane Lands (1620): see 'Reins'
(Great) Reaveley: OE *(ge)refa* = reeve or bailiff. He had general charge of manorial farming. The field is close to 'Manorial Allotment'
(West) Reddis: reeds

PLACE NAMES AND FIELD NAMES OF NORTHUMBERLAND

Redleflate (1498)

Red Peth Shiell Close: reeds or ruddle, probably reed. There is, however, red, redd, riddins = the debris removed from the top of a quarry and the colour red

Red Shaw: wood with reeds?

(The) Reigh

Reins: ON *rein* = a strip of land, boundary strip; reen, reene, reend, rean, reygne, rein = a terraced strip of land on steep hillsides. It means the same as lynchet

Reen in modern use is any division between field strips except unploughed balks. A water channel between rigs is called a reen or open floor

Remelde (1279): OE *helde* = slope

Rester Hill: ME *rest* = resting-place, but it could be from rye

Reverse Knowe-yett (1585) a gate on a hill-top

(The) Rey Hills: OE *ryge* = rye

Resshy Pot (1620): a pit filled with reeds or near to reeds

Riding (1637): OE *rydding* = clearing

(The) Riding

(The New) Rift: ME *rifte* = fissure, cleft. Ploughing

Rigs

Riggy Lea Close (1621)

Rimpeth Dyke (1585): a path that goes round the edge of the land by the dike

Rimpside

Rippitlaw Chesters (1620)

Roaden Hill: near to Rowley Water, from which it may get its name: Roe-deer valley

Roadly Town: either road- or reed- meadow, but OE *rod* = clearing, assart

Roan Tree Field: rowan (rone, roone); a roan, rone is also a clump of gorse (H)

The Roddings: possibly from reeds, but more likely to be from OE *rodum* = (at the) clearings

Rooking: OE *ruh* = rough land

Rosebery Fold (1620)

Ross Dyke (1577): Welsh *rhos* = moor; Irish *ros* = hillock. (In the case of Ros Castle it is both a hill and a moor)

Round Haugh Glebe: the shape of the land by the river

Roundabouts: the field can be encircled by vegetation or water. There can be a circular clump of trees

Rough Deals: dales of rough grassland

Rowledge ford (pre-seventeenth century): OE *ruh* = rough grassland

Rownd Know (1620): a rounded hilltop

(The) Rudds reeds or clearings: OE *hreod* = reed

(the) Rudford (pre-seventeenth century)

Rugley lonying (1567): rough pasture by the track-way

Rumedu (1250): OE *ruh* = rough grassland

Rumpy Face

Rushy Park

Rush Close rushes

Rushey Bog

Ruskie hawuers (1614)

Russel Bank (1620): rushy

Rye Close

Rye Ridds (1620): clearings where rye is grown; OE *ryge* = rye.

Ryley: Rye meadow

Rymessid (1472): at the edge of the sloping land

Rymside (1567)

Saint Foin Haugh: Sainfoin is a leguminous plant used on limestone soils as an improver. In French it means 'healthy hay'

St Helen's Field: dedication

St Mary Knowes (1618)

St Mary Well

St John's Close (1567): The Knights Hospitallers of Mount St John Baptist in Yorkshire held lands at Warkworth

152

Salkelds Moor: Samuel Salkeld owned Fallodon. He published *A New Book of Geography* in 1695. Note also: OE *helde* = a slope. Waste land on a slope where willows grow

Sallywell Closes: OE *salh* = willow

Salmon Field:(a) There have been times when salmon have been taken out of rivers to manure fields, (b) Salmon is a surname, (c) This field lies between Devils Water and Rowley Water

Salt Goats: water outlets into the sea

(The) Salt Grass: a field by the sea

Saltgrese (1471)

Salters Well Field

Saltpan How: the pans or hollows in which salt water was evaporated. Both these fields are on the coast

Sandylands

Sandyeford deane common(1618): common land by the sandy ford valley

Saugh Butts (1623)

Saughey Rig

Saughy Pond Close: OE *salh*, *saugh* = sallow = a willow

Sayning Bank Brege (1585): there is ON *senningr* = dispute, but this is more likely to be from OE *salh* = willow

Sawpit Field: one of the men who operated the two-handed saw had to work below ground level in a special pit

Scallion Hole: (a) a scallion is a young onion in a stage of growth before the bulk has formed (H), (b) ON *skalli* = bald head, bare hill

Scar Field: ON *sker* = rocky outcrop

Scarr Field: scar = a rough, bare precipice

(Newton) Scarthe (1567): scar, rocks or ON *skarth* = pass, gap

(lez) schores (1498): the edges, the coastal land

School Field

Schothalghbankys (1479): furlong of land on the bank of the river

Sclate Burn and Slaitburne (1577): OE *slae* = slope; ON *sletta* = smooth, level fields

(The) Scribe: a unit of land. 1/5 acre

Scroggs Allotment: ME = bush, brushwood

Scurl Hill: Possibly 'rocky' (ON *sker*) or OE *scir* = clear, shining

Sea haughe (1567): a 'heugh' is a cliff, from OE *hoh*

Sea Hewgh pasture (1614)

Sea Lands

Sea Linkes (1567): sand dunes

Seed Field: OE *saed* = area of sown grass

Seggy Hole: OE *secg* = sedge, reed; or it could be clay for saggars

Sedges Moor: sedge, especially the yellow iris

Selbyes foarde (1585): it might be a ford by a willow copse. Selby is a surname

Seldom Seen: a remote field

Senior Croft (1621): a surname

Servant's Close

(The) Seventeen Demain Riggs: 17 riggs, strips of land, belonging to the demesne. (See also the Lords Sixteenth, above)

Sewleyes (1334)

Shanks: the shank of a hill is the projecting part that slopes down to level ground

Sharplaw (1624) OE *scearp* = a steep hill

(Great) Shaw: OE *sceaga* = a copse

(High) Sheath: a parcel of strips of land

Sheep Pasture

Sheepwash: where the sheep are dipped

(Middle) Sheete (1620): A 'sheth' is a group of parallel strips of ploughed land in an open field that adjoin a similar group at right angles

(West) Sheets

(Far) Sheth

Shell Close (1567)

Shelryge (1334): ME *schele* or OE *sele*

Shellythorns (1624): OE *sele* = willow copse

(The) Shepherd's Pasture

(Davy) Shield Common: ME *schele* = a shepherd's summer hut, or small house A 'shieling'

Shildene
Shilling Hill
Shiney Hills
Shining Hall Moor
Shipp Coats (1620): OE *sceap* = sheep
Shipsburn (1624): sheep or *scypen* = cowshed
Shirtnaked Hirds Close: either an indication of poverty, or a mis-copying of 'skirlnaked' – a name given to a hill in the Cheviots, meaning bright, shining
Shittleheugh: OE *scytel-hol* = unstable ridge of land
Shop Field: OE *sceoppa* = shop, shed. A tool store and a lodging
Shoulder of Mutton: the shape of the field
Short Acres: the strips of land in the open-field system are short
Shortrike (1641): short ridge
Shotthaugh: a furlong in the open-field system, by the river
Shovelbred (1235): as narrow as a shovel
Should braids (1624): ME *shovel-brade, brede* = a narrow strip of land (*shoul, shul* = a shovel)
Showlbread (1620) as broad as a shovel
Shrubbery Wood
(Fairney) Sides: OE *side* = hillside, riverside, lakeside. OE *sid* = large, spacious. (Fairney means fern-covered)
Sickett meadow demeane (1614): OE *sic* = small stream, meadow by a stream
Sikett dicke (1585): ON *sik* = ditch, trench
Silloans Common Land
Sills: OE *syle* = bog
Silly Top: covered with willows (salix)
Silly Wrea: nook where willows grew
Silver Spoon Close: the field is shaped like a spoon
Skimmer Pool: perhaps associated with a game or an insect
(North) Slaters: there is no natural slate here, but this may be OE *slaed* = valley, low-lying marsh

Slatt Pitt meadowe (1614)
Slatt Hill (1620): dialect 'slake' means muddy. 'Slack' means hollow
(The) Sledges: (a) OE *slaed* = valley, (b) Sleds and drees were used as transport in addition to wheeled vehicles (H)
Sleeper Law: perhaps the hill field was slow to produce anything, or it could be OE *slaep* = a slippery place
(Great) Sloshes: slosh = sludge, puddle
Slote Allotment: perhaps OE *slah* = sloe
Smalburn Quarter (1621)
Smalburnfeild (1621)
Smale latch rigs (1620): land by the small letch
Small burn (1567)
Small dene
Smartside: ON *sma(r)* = small
Smithy Field
Smoeth meadow (1614): smooth
Smythehopside (1479): Smith's side of the valley
Snabb Butts (1623): projecting hill
Snableases: same place; ME *snabbe*, dialect; *snab* = projecting hill or rock. In this case a steep place
Snapes Leases (1585)
Snapleazes
Snipe Meadow (1620)
Snokoe Fell Top
Snook Farm: a projecting piece of land
Soddy Brig Close: ME *sogh* = soggy, swamp land where there is a bridge
(West) Soldier Close (1770)
Solid Holes
Soppit: boggy land
Sour Plain: OE *sur* = coarse, infertile, acid soil
Soury Close
Southered Riggs: the most southerly land
Southlough Close (1620)
South medow close (1621)
Spartishaw: OE *spearca* = covered with shrubs or brushwood

Spears: perhaps like the above, a meadow covered with brushwood
Sperty medowe (1471)
Spindle Burn Close
Spineles (1620) OE *spinele* = spindle-tree
Spithope Common
Spitle (1620) Hospital of St John the Baptist (Birling)
(Moor) Spott (1770)
Spout Well Close: ME *spoute*: in its earliest form it meant a waterpipe. It is also used in the sense of a small waterfall (e.g. Linhope Spout) or stream
Spreading
Spring Field
Square Field
Stackyard Close
Stafford's Pool
Staith Bank
Stallion Wood Field
(Low) Stamford Lands: probably where the road crossed the stream
Stanecroft (1471): OE *stan* = stone
Stankhill Feild (1621): ME *stank* = pond
(The) Stanners: the stones, in this case a deposit of river gravel at Felton
Staples Allotment Staples Planting: (a) OE *stapol* = pillar, (b) staple, stapple, staple = a well or small pit shaft (H)
Star Myres: (a) 'star grass' is the woodruff. It was dried and put in linen chests because it smelt sweet (H), (b) ON *ston* in compounds means sedge. The mire or bog would produce the right conditions for growth
Station Field: on the railway
Staward Field: stony meadows
Staynge leases (1567)
Stead: OE *stede*, *styde* = place, site
Steele meadowe (1614): OE *stigel* = stile. A meadow with a stile
Stile meadow (1620)
Stiles

Stepping rood Buttes (1620): OE *stybbing* = a clearing
Stepping Stone Lee
Stint Close
Stoats Dale
Stobbs: a 'stob' is a post, stump or stake. Usually the sense is 'stumpy ground'
Stob Hill
(le) Stobithorn (1470)
Stob Ridge Close
Stocks Pasture
Stockwell Riggs
Stokflat (1334): OE *stocc* = place; *stocc* = tree stump, log
Stone Bridge Close
Stone Horse Close
(The) Stoney Field
Stonylands (1577)
(High) Stubbic and (Low) Stubble: stubbly or stumpy ground
Stuffleu Hill and Stifley Hill (1577)
Stumpert lawes (1620): tree stumps
Straithill Letch (1623)
(The) Street Head
(Broad) Street (1620): OE *straet* means street or road: in particular a Roman road when the name is an old one. But OE *steort* = a long projection, certainly in the second example
Streyght-Hills (1479): ? OE *steort* = long projection
(The) Stretch Green: a stretch of grassland, a narrow strip of land
(Quarry) Strip: (where there is a quarry)
(West) Stripe
Stroletch Peece (1620)
Strother: OE *strother* = a place overgrown with brushwood. Also a surname
Suckler Riggs
Sudslyl Butts (1620): short pieces of land on the south side
Sunderlande (1498): OE *sundor-land* = private or separated land

Sunderland flatt (1620)
Sunniside: sunny (south) side
Sunny Close
Swanlawflat (1334): 'Swan' is a surname
Sweeting roods (1614): OE *swete*; ON *eng* = pleasant or fertile meadow land
Swilcher Dean Closes: where the stream washes against the bank
Swinelee (1567): swine pasture
Swinecroft: OE *swin* = swine
Swinleys (1279): OE *swin* = swine
Swinter land hawuers(1618)
Swirlsburn Field: OE *sweorla* = a neck of land
Switchin Moore deane (1620): ON *swithin* = moorland cleared by burning
Swynborne Butts (1620)
Swynhoe Common (1620): the 'hoe' is a ridge of land
Swynburne-feld (1497)
(High) Syke: OE *sic* = a small stream, meadow by a stream. In Northumberland many sikes are feeder streams; ON *sik* = ditch, trench. In the open fields the sike would have provided the unploughed land with drainage
Syket-meadow (1295)

Tank Wood Field
Tanzy Crooks: a herb
Tealy Fens: marshy land with teal (ducks)
Tenacres (1472): ten strips of land
Ten Acre Close
Tenement Meadow Deales(1618): OF, ME *tenement*
Ten Rigs
Tenter Bank (1577): ME *teyntour* = land containing cloth-stretching frames
Tenter Garth Field
Tetherley Balks
Tewhit Plain: the second part of the first word could be ON *thwaite* = clearing
Thatch Dale
Thirlwalls Field: OE *thyrel* = perforated; a gap in the walls

Thistley Moor: thorny land
Thorlow Wood
Thorney dyke lands (1624)
Thorn Hedge Quarter
Threap Moor Manorial Allotment: OE *threap* = in dispute
Throat of the hoop (1620)
Tiallez (1472): the meadow or clearing near which tiles were made
Tilery Field
Tile Shed Close
Tinkler Bank
Todaile medow (1620): OE *todd* = brushwood
Toddailbutts (1620) OE *todd* = fox
Toddle Bush: 'Toddle' is either fox-hole or fox-hill
Tod holes
Toft: an enclosure
Toftor Crofts (1621)
Tongue buttes (1614): OE *tunge* = tongue-shaped land
(The) Tongues
Toplow Bush Close: OE *top* = hill-top; OE *hlaw* = hill
Town Dales (1620): town lands at the end of the village, or belonging to the township
Townfeild flatt (1621)
Tower Close: land next to a tower
Triangular Close
Tuffets Riggs: a tuffet is a clump of grass
Turby Crook: ME *turbarye* = turf pit, a place from which turf is dug
Turnip Close
Tute Close: OE *tot* = a lookout place
Tuttlaw Close (1620)
Twenty-Acre Meadow (1618)
Tylbots Close (1607): could be OE *teag-botl* = a building in a small enclosure

Under Hill (1624): below the hill
Unthancke lands (1618): OE *unthances* = without leave, a squatter's land. It also means ill will, displeasure

Unthank Close
Upron Flatt

(Newton) Villa: a recent name
(le) Vycars Halghes (1585): the vicar's land by the river
(Lez) Vyvers (1471): ME *vivere* = the ponds

Waap Moor: curlew (also whaup, whaap)
Wha Flatt (1624)
Waif Garth
Wainford Rig: ME *wain* = waggon
Walk Mill: OE *walc* = fulling of cloth
(The) Wall Gardens
(Curls) Walls
Wamboys (1495) and Wamobes (1585): OE *hwamm* = a marshy hollow (bois-wood?)
Wander Knowles (1614): ME *wandale, wandel* = a share of land
Wandon Gate (1616)
Wandon Lowfield: it could be like the above, but 'don' is either hill or valley, usually
(Yatesfield) Wanlaces: OE *wang* = meadow; OE *wann* = dark; 'aces' could be 'acres'
Warkneys flatt and deales (1620): OE *walc* = fulling. It is possible that this is where fulling was carried out
Warnfoord In-Moore Comon (1620): ford on the Waren Burn
(Lower) Wash House: OE *waesce* = sheepwash
Washwood bank: OE *waesse* = wet place, swamp
Watch Hill: a lookout place
Water Close (1620)
Water Field
(the) water gait (1580)
gaites (1620)
Wateris Know (1623)
Waterside Close
Waterlees (1472): watermeadows
Waynrig (1477): OE *waeg*; ME *wain* = waggon

(Common) Ways: right of way
(Long) Weasell Flat (1614): OE *wes(u)le* = weasel
Weatifurreans (1614): wet land at the far boundaries
Wedder Pasture: OE *wether* = a castrated ram
Wetenhalghford (1250): a ford at a wet haugh
Well Field: spring
Welsidmedowe (1471): meadow by the spring
Westcrosflat (1334): lands that cross at the west part of the township
Well Way Rigg (1621)
(le) west close (1580)
(Far) Wester Close
Westemestemede (1235): the most westerly meadow
West Reans (1624): western boundary-lands
Westly Bank
West Wham or Bog: OE *hwamm* = corner or valley
(The) Wheat Field
Wheat Holme
Wheatlets
(The) Whimseys: 'whim-gin' is a winding engine worked by horses, and also called a whim or whimsey (H)
Whinney Hill: ME *whin* = gorse, furze (but note that OE *winn* = pasture)
Whip Hill: (a) Whip-grass, (b) a source of stems for making whips (e.g. whitebeam or wayfaring tree)
Whiskers Shields
White Rigs: 'White' is common in field names and describes the colour of various dead grasses and cotton-grass
Whitlees
(Low) Whittridge (1621)
Whitstraye Meadow
Whorlister Wood: OE *hwerfel* = circular enclosure
Whorlton Carre pasture (1616): its present name is 'Whirleyshaws', Qwirlecharr' in 1350 – possibly a quarry wood

Whormesley dyke (1580): it could be 'snakes'
Whorters: OE *waroth* = marshy land by a river
(Hather)wick: OE *wic* = settlement, farm by the heather
Widehoe: in this case a wide haugh, where the stream enters the sea
Wideubb banks: this could be *wilde* = uncultivated, which would make sense from its position on the banks of a stream
Wilie dales (1620): OE *wilig* = willow
Willow Riggs
Windmill Field
Windlass Close
(West) Windy Hill
Windyegg lands (1614): windy edge
Winterside Wood
Wintrick house (1580): OE *winter*; OE *hrycce* = where hay is stacked
(The) Wires: ME *weyour* = a pond; land near a pond
Wisepool close
Wear Haugh: wear, wier = a structure of stone mixed with brushwood to protect the river bank. The haugh has such an embankment
Woodman's Close

Wood Nook
Woodwaies (1620)
Woolaw: probably OE *wulf* = wolf hill
Woumlyhill: (1620) snake-infested? Querns?
Wreigh Hill Law (later Rye Hill): OE *wreo* = twisting (can become 'wry') There is also the possibility of OE *wearg* = a felon
Wretchwood (1620) OE *wrecca-wudu* = outlaw's wood
Wydnewell (1339): well or spring in the wynding (a narrow winding lane)
Wylde mere-mede (1539): meadow on the boundary of uncultivated land

(Broad) Yards: OE *gerd* = a rod in length
Yardside riggs
Yatesfield House: OE *geat* = gate. Field by a gate
Yateside Lonning (1620): a trackway that leads to a gate
Yeal Pool: OE *helde* = sloping land (notice too that Yeeld yows are ewes from which lambs have been weaned)
Year Haugh: Probably from OE *ea* = a stream, as this would fit its position
Yexley: OE *geaces-leah* = cuckoo lea

BIBLIOGRAPHY

Beckensall, S. 1975 *Northumberland Place Names* with two subsequent re-prints (Newcastle)
Beckensall, S. 1977 *Northumberland Field Names* (Newcastle)
Brockett, J.T. 1829 *A Glossary of North Country Words in Use* (Newcastle: Emerson Charnley, Bigg Market; London; Baldwin and Craddock)
Ekwall, E. 1977 *The Concise Oxford Dictionary of Place names* Fourth edition reprinted (OUP)
Gelling, M. 1984 *Place names in the Landscape* (Phoenix Press)
Heslopp, O. 1892 *Northumberland Words* (The English Dialect Society, by Keegan Paul, Trench, Trubner & Co., Charing Cross Road; 1965 Kraus Reprint Ltd., Vaduz)
Mawer, A. 1920 *The Place names of Northumberland and Durham* (CUP)
Northumberland County History Committee 1935. *A History of Northumberland*
Publications of the *English Place name Society*, especially for Cumbria/Cumberland
Watts, V. 2004 The Cambridge Dictionary of English Place names (CUP)

Other books on Northumberland by Stan Beckensall:
2001 *Northumberland: the Power of Place* (Stroud: Tempus)
2001 *Prehistoric Rock Art in Northumberland* (Stroud: Tempus)
2003 *Prehistoric Northumberland* (Stroud: Tempus)
2005 *Northumberland: Shadows of the Past* (Stroud: Tempus)

If you are interested in purchasing other books published by Tempus,
or in case you have difficulty finding any Tempus books in your local bookshop,
you can also place orders directly through our website

www.tempus-publishing.com